Mehlville School District
Individually Focused. Committed to All.

[Toughened Glass]

BRIGHTON STATION, ENGLAND

Brighton Railway Station in Brighton, England, was originally built in 1840. Over the years, people have made many changes to the structure of the station to keep the building safe for train passengers.

In order to make the roof strong, engineers decided to use toughened glass. Toughened glass is glass that has been heated and treated to make it stronger than regular glass. Cast iron beams were also used because they do not rust easily and they provide a lot of support for the heavy panes of glass.

The Brighton Railway Station is a very busy place! It is used daily by 30,000 people. Five hundred trains come through the station each day, including many high-speed trains.

NATIONAL GEOGRAPHIC
SCIENCE

PHYSICAL SCIENCE

NATIONAL
GEOGRAPHIC

School Publishing

PROGRAM AUT

Malcolm B. But

Judith S. Le

Randy Be

Kathy C

David

Program Authors

MALCOLM B. BUTLER, PH.D.

Associate Professor of Science Education,
University of South Florida, St. Petersburg, Florida
SCIENCE

JUDITH SWEENEY LEDERMAN, PH.D.

Director of Teacher Education,
Associate Professor of Science Education,
Department of Mathematics and Science Education,
Illinois Institute of Technology, Chicago, Illinois
SCIENCE

RANDY BELL, PH.D.

Associate Professor of Science Education,
University of Virginia, Charlottesville, Virginia
SCIENCE

KATHY CABE TRUNDLE, PH.D.

Associate Professor of Early Childhood Science
Education, The School of Teaching and Learning,
The Ohio State University, Columbus, Ohio
SCIENCE

DAVID W. MOORE, PH.D.

Professor of Education,
College of Teacher Education and Leadership,
Arizona State University, Tempe, Arizona
LITERACY

Program Reviewers

Amani Abuhabsah
Teacher
Dawes Elementary
Chicago, IL

Maria Aida Alanis, Ph.D.
Elementary Science
Instructional Coordinator
Austin Independent School
District
Austin, TX

Jamillah Bakr
Science Mentor Teacher
Cambridge Public Schools
Cambridge, MA

Gwendolyn Battle-Lavert
Assistant Professor of Education
Indiana Wesleyan University
Marion, IN

Carmen Beadles
Retired Science Instructional
Coach
Dallas Independent School
District
Dallas, TX

Andrea Blake-Garrett, Ed.D.
Science Educational Consultant
Newark, NJ

Lori Bowen
Science Specialist
Fayette County Schools
Lexington, KY

Pamela Breitberg
Lead Science Teacher
Zapata Academy
Chicago, IL

Carol Brueggeman
K–5 Science/Math Resource
Teacher
District 11
Colorado Springs, CO

Miranda Carpenter
Teacher, MS Academy Leader
Imagine School
Bradenton, FL

Samuel Carpenter
Teacher
Coonley Elementary
Chicago, IL

Diane E. Comstock
Science Resource Teacher
Cheyenne Mountain School
District
Colorado Springs, CO

Kelly Culbert
K–5 Science Lab Teacher
Princeton Elementary
Orange County, FL

Program Reviewers continued
on page iv.

Acknowledgments

Grateful acknowledgment is given to
the authors, artists, photographers,
museums, publishers, and agents for
permission to reprint copyrighted
material. Every effort has been made
to secure the appropriate permission.
If any omissions have been made or
if corrections are required, please
contact the Publisher.

Illustrator Credits
All illustrations by Precision Graphics.
All maps by Mapping Specialists.

Photographic Credits
Front Cover Roger Bamber/Alamy
Images.

Credits continue on page EM16.

Neither the Publisher nor the
authors shall be liable for any
damage that may be caused or
sustained or result from conducting
any of the activities in this
publication without specifically
following instructions, undertaking
the activities without proper
supervision, or failing to comply
with the cautions contained herein.

The National Geographic Society
John M. Fahey, Jr.,
President & Chief Executive Officer

Gilbert M. Grosvenor,
Chairman of the Board

Copyright © 2011 The Hampton-
Brown Company, Inc., a wholly
owned subsidiary of the National
Geographic Society, publishing
under the imprints National
Geographic School Publishing and
Hampton-Brown.

National Geographic School Publishing
Hampton-Brown
www.myNGconnect.com

Printed in the USA.
RR Donnelley
Jefferson City, MO

ISBN: 978-0-7362-7770-9

13 14 15 16 17 18 19 20

7 8 9 10

Karri Dawes
K–5 Science Instructional
Support Teacher
Garland Independent
School District
Garland, TX

Richard Day
Science Curriculum Specialist
Union Public Schools
Tulsa, OK

Michele DeMuro
Teacher/Educational
Consultant
Monroe, NY

Richard Ellenburg
Science Lab Teacher
Camelot Elementary
Orlando, FL

Beth Faulkner
Brevard Public Schools
Elementary Training Cadre,
Science Point of Contact,
Teacher, NBCT
Apollo Elementary
Titusville, FL

Kim Feltre
Science Supervisor
Hillsborough School District
Newark, NJ

Judy Fisher
Elementary Curriculum
Coordinator
Virginia Beach Schools
Virginia Beach, VA

Anne Z. Fleming
Teacher
Coonley Elementary
Chicago, IL

Becky Gill, Ed.D.
Principal/Elementary Science
Coordinator
Hough Street Elementary
Barrington, IL

Rebecca Gorinac
Elementary Curriculum
Director
Port Huron Area Schools
Port Huron, MI

Anne Grall Reichel Ed. D.
Educational Leadership/
Curriculum and Instruction
Consultant
Barrington, IL

Mary Haskins, Ph.D.
Professor of Biology
Rockhurst University
Kansas City, MO

Arlene Hayman
Teacher
Paradise Public School District
Las Vegas, NV

DeLene Hoffner
Science Specialist, Science
Methods Professor,
Regis University
Academy 20 School District
Colorado Springs, CO

Cindy Holman
District Science Resource
Teacher
Jefferson County Public
Schools
Louisville, KY

Sarah E. Jesse
Instructional Specialist for
Hands-on Science
Rutherford County Schools
Murfreesboro, TN

Dianne Johnson
Science Curriculum Specialist
Buffalo City School District
Buffalo, NY

Kathleen Jordan
Teacher
Wolf Lake Elementary
Orlando, FL

Renee Kumiega
Teacher
Frontier Central School District
Hamburg, NY

Edel Maeder
K–12 Science Curriculum
Coordinator
Greece Central School District
North Greece, NY

Trish Meegan
Lead Teacher
Coonley Elementary
Chicago, IL

Donna Melpolder
Science Resource Teacher
Chatham County Schools
Chatham, NC

Melissa Mishovsky
Science Lab Teacher
Palmetto Elementary
Orlando, FL

Nancy Moore
Educational Consultant
Port Stanley, Ontario, Canada

Melissa Ray
Teacher
Tyler Run Elementary
Powell, OH

Shelley Reinacher
Science Coach
Auburndale Central
Elementary
Auburndale, FL

Kevin J. Richard
Science Education Consultant,
Office of School Improvement
Michigan Department of
Education
Lansing, MI

Cathe Ritz
Teacher
Louis Agassiz Elementary
Cleveland, OH

Rose Sedely
Science Teacher
Eustis Heights Elementary
Eustis, FL

Robert Sotak, Ed.D.
Science Program Director,
Curriculum and Instruction
Everett Public Schools
Everett, WA

Karen Steele
Teacher
Salt Lake City School District
Salt Lake City, UT

Deborah S. Teuscher
Science Coach and
Planetarium Director
Metropolitan School District
of Pike Township
Indianapolis, IN

Michelle Thrift
Science Instructor
Durrance Elementary
Orlando, FL

Cathy Trent
Teacher
Ft. Myers Beach Elementary
Ft. Myers Beach, FL

Jennifer Turner
Teacher
PS 146
New York, NY

Flavia Valente
Teacher
Oak Hammock Elementary
Port St. Lucie, FL

Deborah Vannatter
District Coach, Science
Specialist
Evansville Vanderburgh School
Corporation
Evansville, IN

Katherine White
Science Coordinator
Milton Hershey School
Hershey, PA

Sandy Yellenberg
Science Coordinator
Santa Clara County Office
of Education
Santa Clara, CA

Hillary Zeune de Soto
Science Strategist
Lunt Elementary
Las Vegas, NV

PHYSICAL SCIENCE

CONTENTS

TECHTREK
myNGconnect.com

Student
eEdition

Vocabulary
Games

Digital
Library

Enrichment
Activities

TECHTREK
myNGconnect.com

Student eEdition | Vocabulary Games | Digital Library | Enrichment Activities

CHAPTER
7

PHYSICAL SCIENCE

What Is Physical Science?

Physical science is the study of the physical world around you. This type of science investigates the properties of different objects, as well as how those objects interact with each other. Physical science includes the study of matter, motion and forces, and many kinds of energy, including light and electricity. People who study how all of these things work together are called physical scientists.

You will learn about these aspects of physical science in this unit:

HOW CAN YOU DESCRIBE AND MEASURE PROPERTIES OF MATTER?

Matter is anything that has mass and takes up space. Physical scientists study all of the different properties of matter. These include size, shape, color, texture, and hardness, as well as mass and volume.

WHAT ARE PHYSICAL AND CHEMICAL CHANGES?

Physical scientists study how matter changes. Sometimes, matter can undergo a physical change in which some of its properties, such as the way it looks, are different, but it stays the same substance. Other times, matter can be chemically changed into an entirely different substance.

HOW DO FORCES ACT?

Physical scientists study how objects move. They also study the forces that act on objects. Some of these forces occur when the objects are touching, but some of them occur even if the objects are not touching, like the force of gravity.

WHAT IS MAGNETISM?

Magnets apply forces to certain objects without touching them. Physical scientists study these forces. They also study magnetic fields, and how the strength of these fields is affected by their distance from the objects they are attracting.

WHAT ARE SOME FORMS OF ENERGY?

Physical scientists study energy in all of its forms. They also learn about how energy can cause changes in the physical world. The different kinds of energy include light, sound, chemical, and mechanical.

WHAT IS SOUND?

Sound is a kind of energy that you can hear. Physical scientists study the properties of sound, including pitch and volume.

WHAT IS ELECTRICITY?

Our world runs on electricity. Electricity is energy that flows through wires. Physical scientists study how electricity flows better through certain objects than others. They also study how it can transform into other forms of energy.

MEET A SCIENTIST

Albert Yu-Min Lin:
Materials Scientist, Digital Archaeologist

Albert Yu-Min Lin is the principal investigator for the Valley of the Khans Project, partially funded by the National Geographic Society. His team's mission is to perform a non-destructive archaeological search for the tomb of Genghis Khan using some of the most advanced digital tools in the world. The team's quest is to identify the location of the tomb without disturbing it, thus maintaining respect and reverence for local Mongolian customs.

As a materials scientist, Albert has applied his expertise in materials characterization to this search that has now recognized him as one of National Geographic Adventure Magazine's 2010 "Adventurers of the Year."

The StarCAVE is the five-walled Cave Automated Virtual Environment, that allows its users to immerse themselves in huge three-dimensional projections.

HOW CAN YOU DESCRIBE AND MEASURE

PROPERTIES OF MATTER?

What can you see in this picture? You can see matter! Everything you can see is made of matter. You can use your senses to describe matter. You can also measure an object to figure out how much matter it contains.

TECHTREK
myNGconnect.com

Student eEdition | Vocabulary Games | Digital Library | Enrichment Activities

The woman's costume is made of matter. The costume is red. Color is just one way to describe matter.

After reading Chapter 1, you will be able to:

- Understand that all objects and substances in the world are made of matter. **PROPERTIES OF MATTER**

- Recognize that matter has properties that can be observed through the senses. **PROPERTIES OF MATTER**

- Classify types of materials or mixtures of substances by using their characteristic properties. **MIXTURES AND SOLUTIONS**

- Observe and describe that mixtures are made by combining solids, liquids, gases, or a combination. **MIXTURES AND SOLUTIONS**

- Identify that water can dissolve some materials. **MIXTURES AND SOLUTIONS**

- Observe that the total mass of a material remains constant whether it is together, in parts, or in a different state. **MEASURING MASS**

- Recognize that matter can be observed or measured with tools such as hand lenses, metric rulers, thermometers, balances, magnets, and graduated cylinders. **MEASURING MASS, MEASURING VOLUME**

- Measure the weight (spring scale) and mass (balances in grams or kilograms) of objects. **MEASURING MASS**

- Describe and compare the volumes of objects using a graduated cylinder. **MEASURING VOLUME**

- Science in a Snap! Recognize that matter has properties that can be observed through the senses. **PROPERTIES OF MATTER**

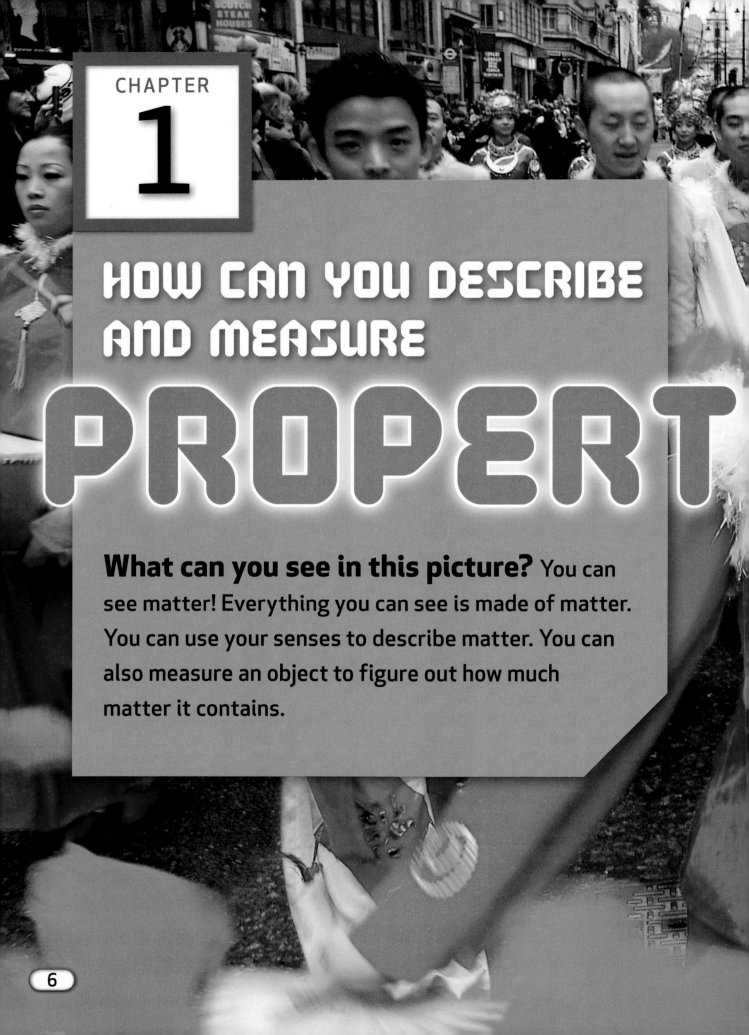

HOW CAN YOU DESCRIBE AND MEASURE

PROPERT

What can you see in this picture? You can see matter! Everything you can see is made of matter. You can use your senses to describe matter. You can also measure an object to figure out how much matter it contains.

The woman's costume is made of matter. The costume is red. Color is just one way to describe matter.

SCIENCE VOCABULARY

matter (MA-ter)

Matter is anything that has mass and takes up space. (p. 10)

This colorful costume is made of matter.

property (PROP-er-tē)

A **property** is something about an object that you can observe with your senses. (p. 10)

Their round shape is one property of these balloons.

mixture (MIKS-chur)

A **mixture** is two or more kinds of matter put together. (p. 18)

A fruit salad is a mixture of solids.

my Science Vocabulary

mass (MAS)	**property** (PROP-er-tē)
matter (MA-ter)	**solution** (so-LŪ-shun)
mixture (MIKS-chur)	**volume** (VOL-yum)

TECHTREK
myNGconnect.com

Vocabulary Games

solution (so-LŪ-shun)

A **solution** is a mixture of two or more kinds of matter evenly spread out. (p. 21)

This juice drink is a solution of water and juice mix.

mass (MAS)

Mass is the amount of matter in an object. (p. 22)

You can use a balance to measure the mass of objects.

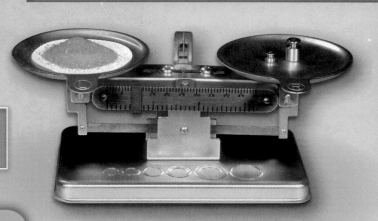

volume (VOL-yum)

Volume is the amount of space something takes up. (p. 26)

The amount of water needed to fill this swimming pool is its volume.

Properties of Matter

Any object or substance is made of **matter**. Matter takes many forms, such as solids, liquids, and gases. You can describe matter by talking about its properties. A **property** is a trait or a quality of a material that we can observe. Look at the hot air balloons. They all have about the same shape. But they have different colors and sizes. No two pieces of matter can be in the same place at the same time. That's why each of these balloons has its own space in the sky.

SOME PROPERTIES WE CAN USE TO DESCRIBE MATTER ARE:

- Shape
- Color
- Odor
- Taste
- Texture
- Hardness
- Attraction to magnets

You could use the properties of shape, size, and color to describe this hot air balloon.

Shape and Color We can observe the shape and color of something. Depending on how you use an object, shape can be very important. Rounded objects can roll, while flat-sided objects will stay put. Shape is important when things need to fit together in order to work.

Color can also tell about an object. Look at the picture of red and black coals. You know the red coals are too hot to touch. Your eyes see different colors. Think of a rainbow. Natural objects, such as flowers, come in many different colors. The things people make, such as clothing and cars, also come in different colors.

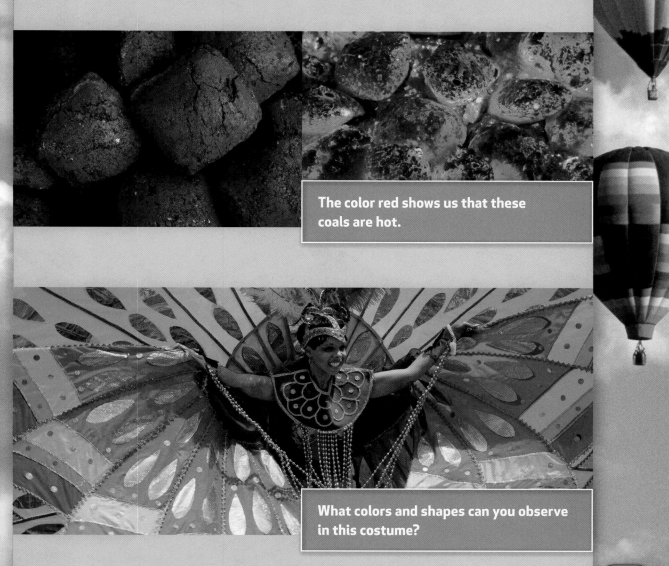

The color red shows us that these coals are hot.

What colors and shapes can you observe in this costume?

Odor and Taste   Many substances have different smells, or odors. Think about the odor of a smoky campfire, a fresh-baked loaf of bread, a flower, or even a container of sour milk. The odor of the bread makes you hungry! And the odor of the milk lets you know that you should not drink that milk.

You use your tongue to taste. Some things taste bitter, while others are salty. Some things taste sweet, and others are sour.

Each type of dried fruit has its own taste.

Science in a Snap! Recognizing Smells

Working in pairs, add a small amount of each substance to a cup. Label the cups and cover them with cling wrap.

One student, the Tester, is blindfolded. Have the Tester smell each scent. Can the Tester recognize the scents?

Which substances were easier or more difficult to recognize?

You use your nose to detect odors. Different flowers have different odors.

Texture and Hardness If you rubbed your hand over a piece of sandpaper, it would feel different than rubbing your hand over a smooth stone. The way an object feels is its texture. Some words that describe texture are *smooth*, *bumpy*, and *rough*. Texture is a property of matter. The table below shows the textures of some types of matter. These close-up views of matter can help us classify them.

TEXTURES OF MATERIALS

MATERIAL

PAPER	SMOOTH	ROUGH
PLASTIC	SMOOTH	BUMPY
METAL	SMOOTH	ROUGH
GLASS	SMOOTH	BUMPY

Hardness is another property of matter. If you put your finger on a mirror, for example, you can't push your finger into the mirror. The mirror is a hard surface. If you put your finger on a pillow, your finger would sink into it. A pillow is soft. One of the hardest substances is a diamond. Knife blades made of diamond can cut through rocks.

Both the graphite in pencil lead and diamonds are made of carbon. But pencil lead is soft, and diamonds are hard.

Which objects in this photo are hard? Which are soft?

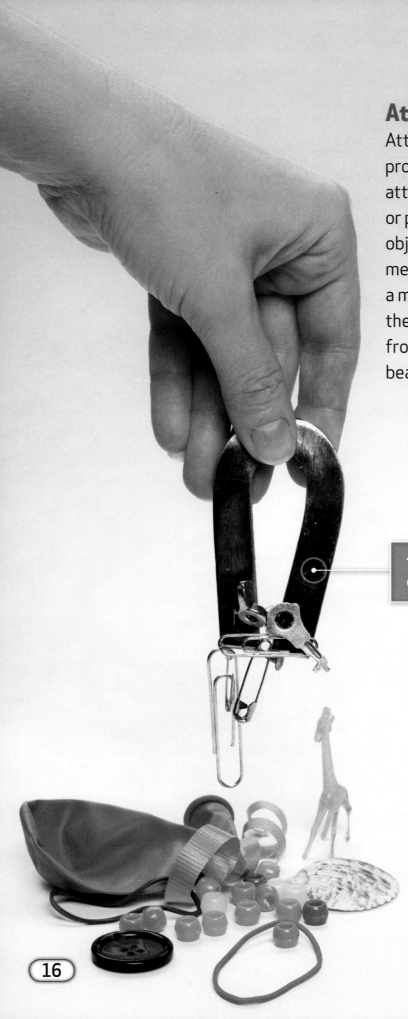

Attraction to Magnets

Attraction to magnets is another property of matter. Magnets do not attract objects such as wood, glass, or pieces of fabric. Magnets attract objects that are made of some metals, such as iron. You could use a magnet, for example, to separate the paper clips and metal objects from the nonmetal objects such as beads, rubber bands, and ribbons.

This magnet attracts and picks up objects made of some kinds of metal.

Because magnets have such unique properties, we can use them in many ways. Look around your house and classroom. What items use magnets? What items do magnets attract?

USES OF MAGNETS

TECHTREK
myNGconnect.com

Enrichment
Activities

If you have used a compass to find your way, you have used a magnet. Compass needles always point north because of magnets.

Computer hard drives use magnets to store information.

Magnets can be handy objects to have around the house!

Before You Move On

1. What properties can you use to describe matter?
2. Compare properties of a glass drinking cup and a metal paper clip. Create a Venn diagram of your observations.
3. **Apply** Which properties of matter can you use to describe a pencil?

Mixtures and Solutions

Look at the picture of the fruit salad. The pieces of fruit are a **mixture** . A mixture is made of two or more things that can be separated from one another. You could separate the pieces of fruit according to their sizes, shapes, and colors.

Mixtures are made of different substances. They do not have to have equal amounts of the substances. Each part of the mixture has its own properties. You can separate a mixture based on the properties of the substances in it.

The pieces of fruit in this mixture are different sizes, colors, and shapes.

Mixtures can be made of solids, liquids, gases, or a combination of these. Look at the photo of the orange juice. The juice is a liquid, and the pulp is a solid. You could separate the juice by pouring it through a strainer, leaving the seeds behind.

Some mixtures, such as oil and vinegar salad dressing or gasoline, are made of only liquids. Other mixtures are made of only solid matter. The trail mix, for example, is a mixture with all solid parts.

This fresh juice is a mixture of solids and a liquid.

This trail mix has all solid parts.

What are some other ways to separate mixtures? Dirt is a mixture of solids. It includes soil, sand, and tiny bits of gravel. You could separate dirt into its different parts by dropping it into a screen that lets different sizes of matter fall through it.

Take a look at the mixture of paper clips. They are all the same shape. So how could you separate them? You might sort the paper clips by their sizes and colors. What if the paper clips were mixed with other objects, such as rocks and marbles? Then you could use a magnet to pull the paper clips from the mixture.

What properties would you use to sort these paper clips?

When you put together the items in a mixture, you can separate them. But what happens if you mix juice powder in water? The juice powder disappears. The juice you mixed is a **solution**. A solution is a mixture of two or more types of matter that are evenly spread out and not easily separated.

Water makes juice mix dissolve to create a solution. But not everything can dissolve in water. If you mixed sugar in water, you'd get a solution. The sugar would dissolve! But what if you put sand in water? Then you would have a mixture of a solid, sand, and a liquid, water.

Juice mix was dissolved in water to make these solutions.

Before You Move On

1. What is a mixture? Name a mixture made of only solids.
2. You have a mixture of feathers, blocks, rubber bands, and paper clips. How might you separate this mixture?
3. **Analyze** Explain the difference between a mixture and a solution. Give examples to explain.

Measuring Mass

If you want to figure out the amount of matter in an object, you can measure its mass . Mass is the amount of matter in an object or substance. Mass is measured in grams and kilograms.

You might step on a scale to weigh yourself. When you step on the scale, you are measuring your weight, not your mass. Weight is measured in pounds. Mass and weight are not the same. Weight depends on the pull of gravity. An astronaut, for example, weighs less on the moon than on Earth. Why? The pull of gravity is stronger on Earth than on the moon. The mass of the astronaut, however, is the same in both places. Only the weight changes.

The astronaut weighs less on the moon because gravity on the moon is weaker than on Earth. His mass remains the same in both places.

You can use a two-pan balance to measure mass. Place the object on one pan. Place mass pieces on the other pan. When the pans balance, you will know the mass of the object. A two-pan balance also can help you determine which of two objects has the greater mass.

A two-pan balance measures mass using standards.

You can use a spring scale to measure the weight of an object.

You can measure the mass of a liquid as well as the mass of a solid. To measure the mass of a liquid, measure the mass of the container. Add the liquid and measure the total mass of the liquid and container. Then subtract the mass of the container to find the liquid's mass.

MEASURING MASS OF A LIQUID

Measure the mass of the container before adding the liquid. Add the liquid.

Measure the mass of the container plus the liquid. Subtract the first measurement from the second. The result is the mass of the liquid.

An object has the same mass as the total mass of all of its parts. Objects don't lose mass if they are broken into parts. Scientists explain this with the Law of Conservation of Mass. If you cut an apple in half, for example, the two halves would have the same mass as the entire apple. Cutting the apple won't change its total mass.

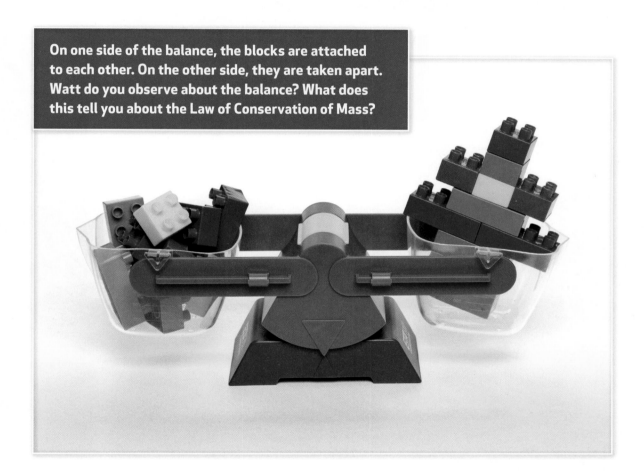

On one side of the balance, the blocks are attached to each other. On the other side, they are taken apart. Watt do you observe about the balance? What does this tell you about the Law of Conservation of Mass?

Before You Move On

1. Explain how mass and weight are different from each other.
2. Identify which two measurements are needed to find the mass of a liquid.
3. **Apply** Design a plan to determine the mass of a puppy that will not sit still on a balance.

Measuring Volume

Mass is the amount of matter in an object. **Volume** is the amount of space that matter takes up. You use volume, for example, in recipes. If you did not add the correct volume of liquid, your recipe would not turn out correctly! The metric unit of liquid volume is the liter. You can also measure volume in fluid ounces.

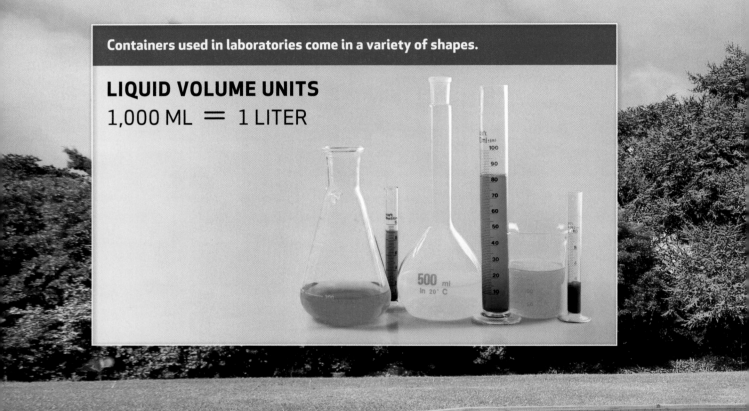

Containers used in laboratories come in a variety of shapes.

LIQUID VOLUME UNITS
1,000 ML = 1 LITER

The amount of water needed to fill this pool is its volume.

Graduated cylinders are good tools for measuring volume. They have lines to show the amount of liquid. You might think a liquid has a flat top. But look closely. You'll see it is slightly curved at the top.

Be sure to read the measurement at eye level and at the bottom of the curve.

How is measuring the volume of a solid different from measuring the volume of a liquid? You can use math to measure the volumes of many solids. If you have a rectangular block, you can measure its length, width, and height. Then you multiply those measurements together. Solid volumes are expressed in cubic centimeters (cm^3) or cubic meters (m^3).

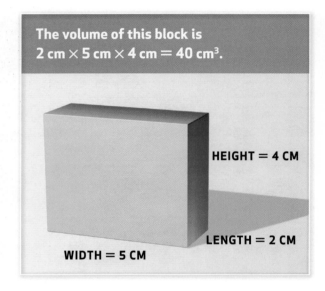

The volume of this block is
$2\ cm \times 5\ cm \times 4\ cm = 40\ cm^3$.

HEIGHT = 4 CM

LENGTH = 2 CM

WIDTH = 5 CM

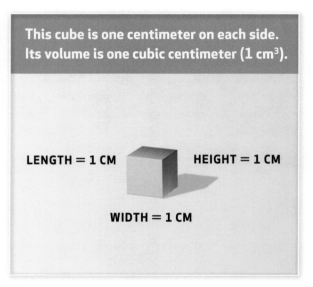

This cube is one centimeter on each side. Its volume is one cubic centimeter ($1\ cm^3$).

LENGTH = 1 CM   HEIGHT = 1 CM

WIDTH = 1 CM

How would you measure the volume of each of these blocks?

You can use a ruler to measure the volume of some solid objects. But what if the solid object doesn't have straight edges? You can measure its volume in another way. Put liquid in a beaker and measure its volume. Then add the object to the beaker and measure the volume again. If you subtract the first volume from the second volume, you'll know the volume of the solid object.

Placing the object in water moves the water. The amount that the water moves is the volume of the object.

Before You Move On

1. Explain how you would measure the volume of a rectangular block.
2. You have found a rock and want to measure its volume. What might you do to measure the volume?
3. **Infer** Can you determine from looking at a container the volume of a liquid it could hold? Explain why or why not.

AVALANCHE SAFETY

Do you like snow? Skiers, snowmobilers, and snowboarders love it! Snow can be lots of fun. But it can also be dangerous. On a mountain, an avalanche of snow can be deadly.

An avalanche happens when a slab of snow slides off the layers beneath. In five seconds, an avalanche can reach 129 kilometers per hour (80 mph)! Fortunately there are experts such as Doug Driskell. He is a Snow Safety Coordinator in Aspen, Colorado.

An avalanche in Alaska

In New Hampshire, experts study the snow layers to determine stability.

Driskell and his coworkers examine the properties of snow. They take many measurements. This helps them predict possible avalanches. Each layer of snow has its own properties. For instance, rain or wet snow will make a layer hard and icy. The mass and volume of snowfall are also considered.

Driskell and his crew try to stop avalanches from happening. There are a few ways to do this. They close dangerous areas off to the public. Sometimes they try to press the snow tightly together. Other times, they might use explosives to create a controlled avalanche. Their goal is to let people enjoy the snow-covered mountains safely.

An avalanche safety worker digs deep to expose the layers of snow. The properties of the layers will be analyzed to see if conditions are right for an avalanche.

Conclusion

Everything is made of matter. You can describe matter by observing its properties, such as shape, color, taste, odor, texture, and hardness. Some matter is attracted to magnets. You can measure an object's mass and volume to describe its size. Mass is the amount of matter in an object, while volume is the amount of space that matter takes up. Mixtures and solutions are both made of two or more kinds of matter. Mixtures can be easily separated, while solutions cannot be easily separated.

Big Idea Matter can be described by its properties and measured by mass and volume.

SOME PROPERTIES OF MATTER

Color

Hardness

Texture

Vocabulary Review

Match the following terms with the correct definition.

A. matter	**1.** Something about an object that you can observe with your senses
B. property	
C. mixture	**2.** Anything that has mass and takes up space
D. solution	**3.** Two or more kinds of matter put together
E. mass	**4.** The amount of space something takes up
F. volume	**5.** The amount of matter in an object
	6. Two or more kinds of matter evenly spread out

Big Idea Review

1. **Explain** What property can you observe with your nose?

2. **Identify** What tool can you use to measure mass?

3. **Calculate** Determine the volume of a block that is 4 cm long, 7 cm wide, and 3 cm high.

4. **Analyze** Imagine that you were describing the properties of a lemon. What properties do you think are most important in describing lemons? Why do you think so?

5. **Explain** Could you measure the volume of a block with straight edges by measuring liquid in a beaker, placing the object in the water, and measuring the volume again? Or does this method only work for objects without straight edges? Explain your answer.

6. **Think Critically** Suppose you need to measure the volume of water needed to fill an aquarium. What would you do?

my
SCIENCE
notebook

Write About How to Describe and Measure Matter

Explain What is happening in this picture? What does the picture tell us about describing matter?

CHAPTER 1

PHYSICAL SCIENCE EXPERT: MECHANICAL ENGINEER

"I was very curious as a youth," says Dr. Albert Yu-Min Lin. "I wanted to explore places." Today, he works as a scientist. Dr. Lin studied and trained as a mechanical engineer. He specializes in understanding the properties of materials.

Dr. Albert Lin is a mechanical engineer at University of California at San Diego.

Dr. Lin often studies natural materials. He thinks of ways these materials can be useful to people. For example, police need body armor. This armor must be made out of a strong material. Dr. Lin thought of the hard shell of the abalone. He and Dr. Marc Meyers are working together on this problem.

Abalone has a tough shell.

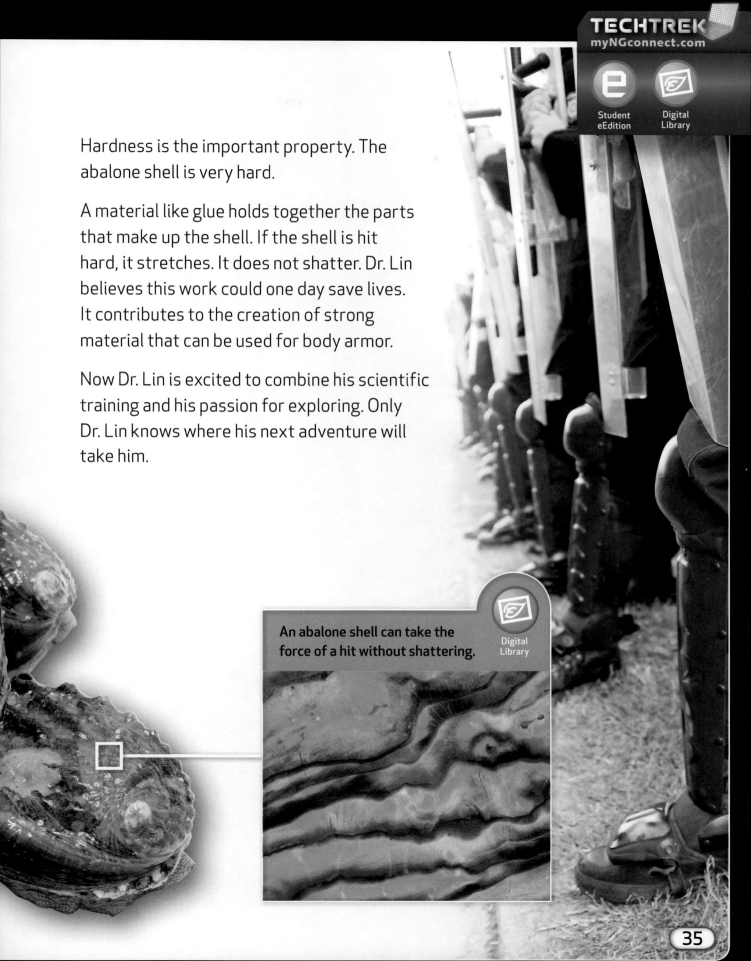

TECHTREK
myNGconnect.com

Student
eEdition

Digital
Library

Hardness is the important property. The abalone shell is very hard.

A material like glue holds together the parts that make up the shell. If the shell is hit hard, it stretches. It does not shatter. Dr. Lin believes this work could one day save lives. It contributes to the creation of strong material that can be used for body armor.

Now Dr. Lin is excited to combine his scientific training and his passion for exploring. Only Dr. Lin knows where his next adventure will take him.

An abalone shell can take the force of a hit without shattering.

Digital
Library

BECOME AN EXPERT

Searching for the Tomb of Genghis Khan: Exploring Remote Mongolia

Genghis Khan ruled a mighty empire before he died in 1227. Even today, people do not know how and where he was buried. Some people believe that horses were used to trample over the area so no one could find the grave. Others believe that soldiers changed the direction of a river to make the area flood. No one has found the tomb in 800 years.

TECHTREK
myNGconnect.com

Digital Library

Genghis Khan conquered more land than any other person in history. His loyal subjects kept the secret of his burial site a mystery for centuries.

TECHTREK
myNGconnect.com

Student
eEdition

Digital
Library

You've already found out that Dr. Albert Lin creates strong **solutions** from various materials to make useful things such as body armor. Dr. Lin is also interested in the mysteries of ancient history. He has led a search to find the tomb of Genghis Khan. He believes it is in a remote, mountainous area of northern Mongolia. The area was guarded for hundreds of years by a tribe. Tribe members would kill anyone who entered the area!

The team cannot simply start digging. That would not show respect for the burial site. Instead of shovels, they must use other tools. They will search for the tomb using satellites and computers.

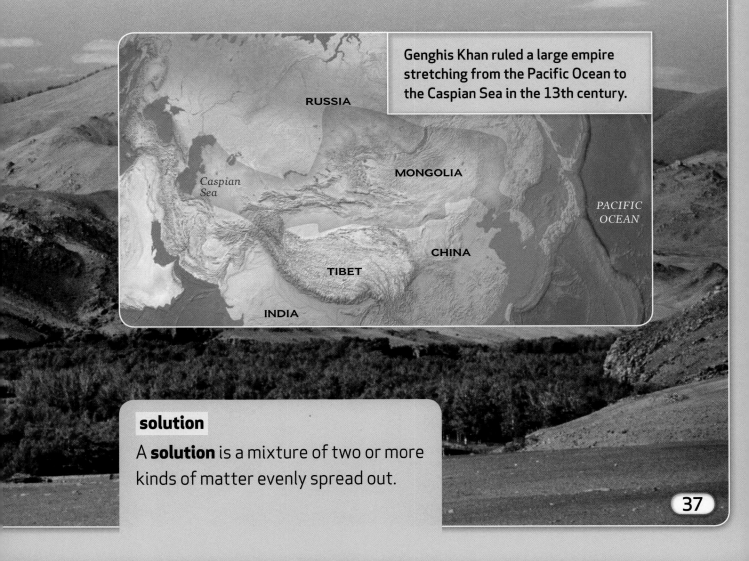

Genghis Khan ruled a large empire stretching from the Pacific Ocean to the Caspian Sea in the 13th century.

RUSSIA

Caspian Sea

MONGOLIA

PACIFIC OCEAN

CHINA

TIBET

INDIA

solution

A **solution** is a mixture of two or more kinds of matter evenly spread out.

Satellite Imaging The team starts by making a map. They use pictures taken by a satellite. The satellite orbits Earth. Mongolian stories told about mountains, rivers, and valleys that could be near the burial site. These important clues could lead the team to the burial site.

This is what a satellite image looks like.

Cell phones, Global Positioning Systems (GPS), and weather prediction all depend on man-made satellites.

Monitoring Properties   The team examines the images closely. They are looking for **properties** of **matter** that might not exist in nature. They search for shapes, colors, and textures. For example, a straight line might suggest a buried wall. A square might suggest a building. These properties are clues that might indicate where the tomb is hidden.

A possible structure made by people, as seen by aerial photography.

matter

Matter is anything that has mass and takes up space.

property

A **property** is something about an object that you can observe with your senses.

The "Eye" of the Mountain Perched on one mountaintop is the "Eye." It is a huge sea of rocks, a **mixture** of rocks of different shapes and sizes. Local people believe that this rock formation could be the tomb of Genghis Khan. Dr. Lin is not so sure. Perhaps it is just a natural rock formation. Was it created by people or not?

The sea of rock called the "Eye" sits on top of a mountain. It is so vast it can be seen by satellite images. Is it natural or made by people?

mixture

A **mixture** is two or more kinds of matter put together.

To solve this riddle scientifically, the team collects samples and measurements. The dimensions of the pile of rock are measured. From those, the **volume** of the pile is estimated. The team measures the **mass** of the rocks. Most of the data makes Dr. Lin believe this is a natural rock formation.

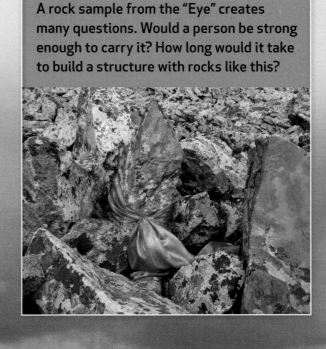

A rock sample from the "Eye" creates many questions. Would a person be strong enough to carry it? How long would it take to build a structure with rocks like this?

volume

Volume is the amount of space something takes up.

mass

Mass is the amount of matter in an object.

Team Work Dr. Lin hopes the public can help search his images. The team wants to create a computer program. This program would let anyone look at the images. Imagine thousands of helpers looking for unnatural patterns. The job would become easier. Even school students could help. They could spend a few minutes looking through photos.

The final phase of the project will start when a possible site is found. Different tools will be used. These tools will not disturb the burial site.

Many thousands of images must be analyzed to search for the tomb.

The search for Genghis Khan's tomb is a big project. It uses a range of tools to describe and measure the matter of the Mongolian landscape. Dr. Lin's goal is to find the tomb. He doesn't want to dig it up. He wants to find the site and protect it. The tomb is an important piece of world history.

As Dr. Lin says, "Our modern world wasn't only created by explorers like Marco Polo and Columbus. It was also created by people from Asia. To use science to show that is an honor for me."

"I hope that the work we do here will not only revive the name of Genghis Khan, but also the discussion about his impact on the world," says Dr. Lin.

SHARE AND COMPARE

Turn and Talk How does Dr. Lin use properties of matter in his work? Form a complete answer to this question together with a partner.

Read Select two pages in this section. Practice reading the pages. Then read aloud to a partner. Talk about why the pages are interesting.

Write Write a conclusion that tells the important ideas you learned about the way in which researchers use properties of matter to find important sites from history. State what you think is the Big Idea of this section. Share what you wrote with a classmate. Compare your conclusions. Did your classmate make the same connection between the properties of matter and their usefulness for research?

Draw Imagine that Dr. Lin's team finds the tomb of Genghis Khan. What properties of matter will the research team observe? Draw what you think they might find. Combine your drawings with those of your classmates to make a "field diary."

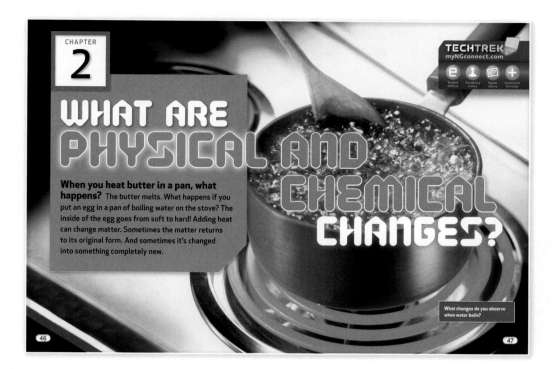

CHAPTER
2

WHAT ARE PHYSICAL AND CHEMICAL CHANGES?

When you heat butter in a pan, what happens? The butter melts. What happens if you put an egg in a pan of boiling water on the stove? The inside of the egg goes from soft to hard! Adding heat can change matter. Sometimes the matter returns to its original form. And sometimes it's changed into something completely new.

TECHTREK
myNGconnect.com

What changes do you observe when water boils?

46   47

After reading Chapter 2, you will be able to:

- Compare and contrast the states of matter. **STATES OF MATTER**

- Explain how matter can change from one state to another by heating and cooling.
  STATES OF MATTER

- Realize that water has unique properties that make it an important resource.
  STATES OF MATTER

- Understand that, when a new material is made by combining two or more materials, it has properties that are different from the original materials.
  CHANGES THAT MAKE NEW MATTER

- Observe and describe changes in the properties of materials or objects.
  CHANGES THAT MAKE NEW MATTER

- Science in a Snap! Observe and describe changes in the properties of materials or objects.
  CHANGES THAT MAKE NEW MATTER

WHAT ARE PHYSICAL

When you heat butter in a pan, what happens? The butter melts. What happens if you put an egg in a pan of boiling water on the stove? The inside of the egg goes from soft to hard! Adding heat can change matter. Sometimes the matter returns to its original form. And sometimes it's changed into something completely new.

TECHTREK
myNGconnect.com

Student
eEdition

Vocabulary
Games

Digital
Library

Enrichment
Activities

AND CHEMICAL CHANGES?

What changes do you observe when water boils?

SCIENCE VOCABULARY

states of matter
(STĀTS UV MA-ter)

States of matter are the forms in which a material can exist. (p. 50)

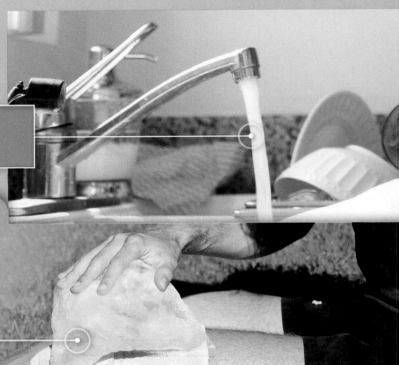

Liquid is one of the states of matter in which water exists.

solid (SO-lid)

A **solid** is matter that has a definite shape and volume. (p. 50)

Ice is a solid.

physical change
(FI-si-kul chānj)

A **physical change** is when matter changes to look different but does not become a new kind of matter. (p. 51)

The change from liquid water to ice is a physical change.

my
Science
Vocabulary

chemical change
(KEM-i-kul chānj)

gas
(GAS)

liquid
(LI-kwid)

physical change
(FI-si-kul chānj)

solid
(SO-lid)

states of matter
(STĀTS UV MA-ter)

TECHTREK
myNGconnect.com

Vocabulary
Games

liquid (LI-kwid)

A **liquid** is matter that has a definite volume and takes the shape of its container. (p. 52)

Water is a liquid.

gas (GAS)

A **gas** is matter that spreads to fill a space. (p. 54)

The humidifier, using the gas form of water, helps keep the air moist.

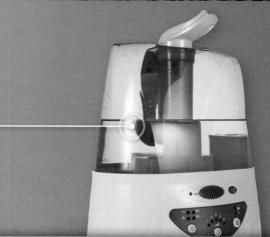

chemical change
(KEM-i-kul chānj)

A **chemical change** is a change in which new substances are formed. (p. 56)

Decay is one kind of chemical change.

49

States of Matter

Water as a Solid Take a look at the photograph on this page. The driver is moving a large machine over an ice rink. Ice is one state, or form, of water. The **states of matter** are simply the forms that matter can take.

Ice is water in its **solid** state. A solid is matter that has a definite shape and volume. Think about an ice cube. If you moved it from one container to another, it would stay the same shape and size. This book, your desk, and your school building are all solids too.

This machine sprays water onto the ice. The water quickly freezes. The result is a fresh, smooth surface for ice skaters or hockey players.

What makes water turn into its solid state? The change from water to ice is caused by a change in temperature. When the temperature of water drops to 0°C (32°F), water freezes and becomes solid.

The change of water from a liquid to a solid is called a **physical change**. When the water changes from liquid to ice, it is still water. It has simply changed state because of temperature changes.

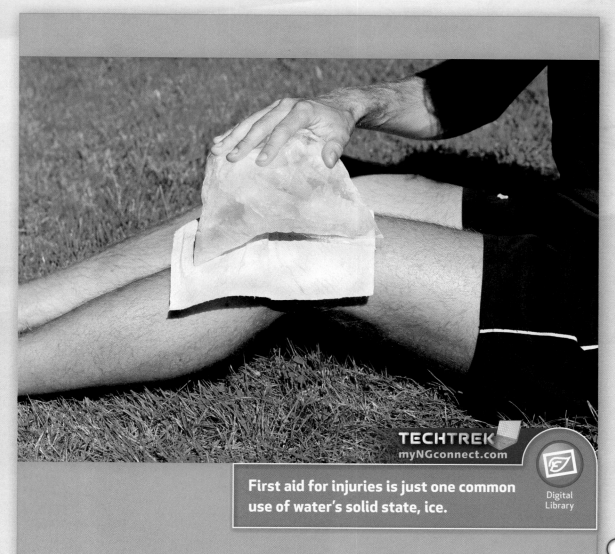

First aid for injuries is just one common use of water's solid state, ice.

Water as a Liquid Imagine you have a cup of ice. You set it out in the warm sun. What happens? The ice undergoes a physical change. It melts. When the temperature of ice rises above 0°C (32°F), solid ice begins to turn into its **liquid** state. A liquid is matter that has a definite volume and takes the shape of its container. If you had water in a cup and poured it into a bowl, you'd have the same volume, or amount, of water. It would just change from the shape of the cup to the shape of the bowl.

Water is the most common liquid in the world. Animals and plants need water to live. People use water for many things, including drinking, cleaning, and bathing.

TECHTREK
myNGconnect.com

Digital
Library

Water in its liquid form helps living things grow.

Washing is only one of the many ways people use water every day.

Water as a Gas Have you ever blown up a balloon? As you blow into a balloon, the sides get bigger and bigger. That's because the air that you blow into the balloon is a **gas** . A gas is a type of matter that spreads out to fill a space.

Water, in the form of a gas, helps this train run.

When the temperature of liquid water rises to 100°C (212°F), liquid water begins to change state. It turns into a gas called water vapor. You cannot see water vapor in the air. The water vapor is invisible. Water vapor in the air helps keep the air from getting too dry. People use humidifiers to add water vapor to the air.

The humidifier helps keep the air moist.

Before You Move On

1. What are the three states, or forms, that water can take?
2. Choose six objects or materials in the room. Classify them. Which are solids? Liquids? Gases?
3. **Apply** Describe the physical change that happens if you put ice cubes in a glass and let the glass sit at room temperature.

Changes That Make New Matter

Decaying When temperatures drop low enough, liquid water turns to ice. If the temperature rises, solid ice changes back to its liquid state, water. In its liquid, gas, or solid form, water is still water. It looks different, but it is still the same kind of matter. The water undergoes a physical change.

Sometimes when matter changes, though, the change is permanent. The matter cannot go back to how it was before. These permanent changes are called **chemical changes** . Take a look at the apple in the photograph. When an apple is cut and left out in the air, its fleshy part turns brown. The apple rots, or decays. The decaying of food is a chemical change.

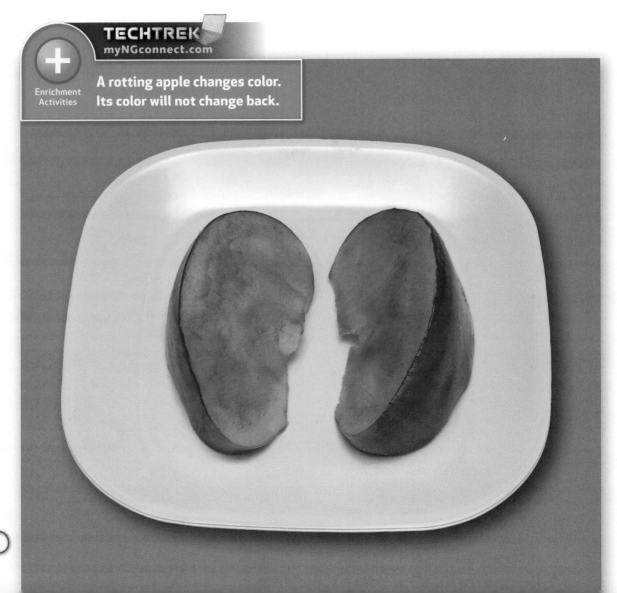

TECHTREK
myNGconnect.com

Enrichment
Activities

A rotting apple changes color. Its color will not change back.

Apples are not the only natural things that decay. This fish decayed, too. Dead plants and animals decay over time. As living things decay, they change color to black or brown. They give off a bad odor, or smell. The change in color and smell are signs of a chemical change. Once an animal or plant has decayed, the change is permanent.

This fish died and washed up on the shore. How has the fish changed as it decayed?

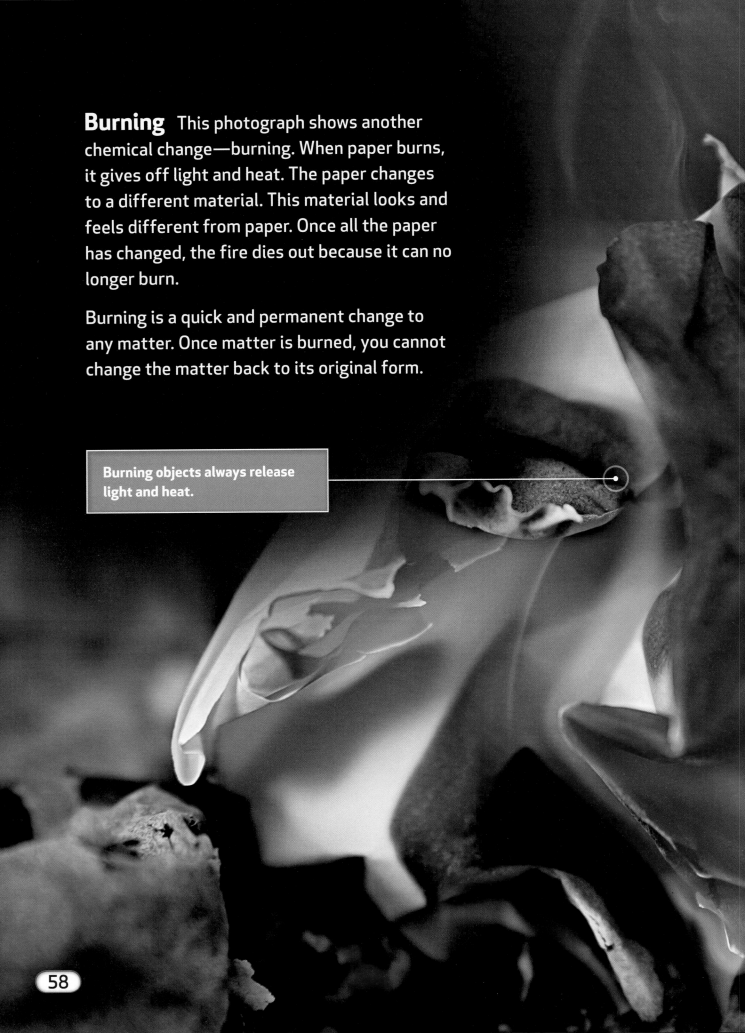

Burning This photograph shows another chemical change—burning. When paper burns, it gives off light and heat. The paper changes to a different material. This material looks and feels different from paper. Once all the paper has changed, the fire dies out because it can no longer burn.

Burning is a quick and permanent change to any matter. Once matter is burned, you cannot change the matter back to its original form.

Burning objects always release light and heat.

Rusting

Iron is a strong gray metal. But iron can change to rust—a brittle orange solid. Rusting is a chemical change that happens when iron combines with oxygen. Rusting needs water to happen too.

Because iron is so strong, people use it to make steel. Steel is used in cars, ships, and buildings. But many steel things are outside in the air. If rain falls on these objects, rust can form. So people often paint or coat objects made out of steel to protect them from changing to rust.

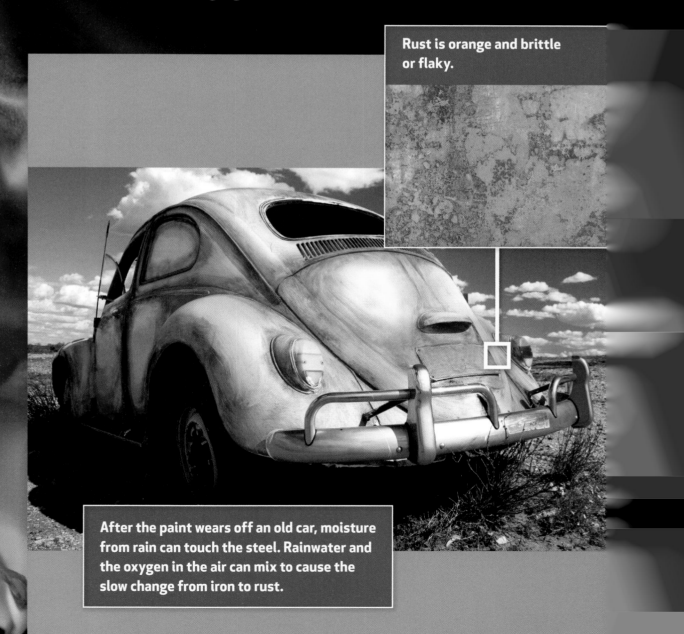

Rust is orange and brittle or flaky.

After the paint wears off an old car, moisture from rain can touch the steel. Rainwater and the oxygen in the air can mix to cause the slow change from iron to rust.

Cooking and Baking Cooking and baking are two delicious ways in which matter can change! You can brown a piece of meat by heating it. Once the meat is browned, the change is permanent—the meat cannot change back to what it was like before it was cooked. When you cook an egg, all or part of the egg changes from a liquid to a solid.

Baking is a chemical change too. You can mix butter, flour, and other ingredients and change them into cake or bread by baking them. When you bake bread, the yeast gives off a gas that makes the bread rise. The smell of the batter changes when it bakes. The dough changes color too. The changes in smell and color and the forming of gas are all signs of the change in matter that happens during baking.

What changes can you observe when dough is baked to form bread?

60

Pour 50 mL of vinegar into a small plastic bottle. Put 2 spoonfuls of baking soda into a balloon.

Carefully stretch the opening of the balloon over the opening of the bottle without letting the baking soda fall into the bottle. Tip the balloon so that the baking soda goes into the bottle.

What did you observe? How did matter change?

Before You Move On

1. Name at least two kinds of chemical changes.
2. A raw meatball is put into a hot pan. What kinds of changes do you think you will observe?
3. **Compare** Water can change to ice. A liquid egg can change to a solid egg. How are these changes the same? How are they different?

RUSTING METAL
IN SUNKEN SHIPS

Rust forms when oxygen and iron combine when water is present. Oxygen is a gas. It makes up about 21 percent of the air. Iron things that are surrounded by air and exposed to moisture will rust.

But sunken ships also rust. How? Things rust underwater because oxygen dissolves in water. There is more oxygen in shallow water than in deep water. So, sunken ships that are near the surface of the ocean rust more quickly than ships that are deep underwater.

Lying in shallow water where there is plenty of oxygen, this ship's iron is well rusted.

The wreck of *Titanic* is about 4 km (about 2.5 miles) underwater. Because it is so far underwater, scientists thought that it would not rust quickly. But they discovered that *Titanic* is rusting in a different way. Iron-eating bacteria are making the ship rust! The bacteria make "rusticles" on *Titanic*. Rusticles look like icicles, but are made of rust and bacteria. The bacteria that make rusticles live only deep underwater. No one had ever seen them before people discovered the wreck of *Titanic*.

Normal rust does not change the shape of a sunken ship, but rusticles make a ship look like it's covered in dripping icicles.

Water exists in three forms, or states: solid, liquid, and gas. Water changes from one state to another because of changes in temperature. These changes are called physical changes. Matter can change in other ways too. Decaying, rusting, burning, and cooking are chemical changes. The changes they cause are permanent.

Big Idea Matter can change in many ways.

WAYS THAT MATTER CAN CHANGE

Decaying	Burning	Rusting	Cooking

Vocabulary Review

Match the following terms with the correct definition.

A. states of matter

B. solid

C. physical change

D. liquid

E. gas

F. chemical change

1. Matter that spreads to fill a space
2. The forms in which a material can exist
3. Matter that has a definite shape and volume
4. Matter that has a definite volume and takes the shape of its container
5. A change in which matter changes to look different but does not become a new kind of matter
6. A change in which new substances are formed

Big Idea Review

1. **Explain** What changes can you observe when iron changes to rust?

2. **Define** What is a gas?

3. **Identify** Identify a change in matter that you can see in your home. Is the change a physical change or a chemical change?

4. **Compare** Liquid water becomes water vapor. Paper burns and changes. How are these changes the same? How are they different?

5. **Infer** Are burning, decaying, and cooking changes of state? Explain why or why not.

6. **Apply** You are helping someone cook a meal. Describe the changes that take place in the ingredients or the foods. Tell what signs of change you can observe in the foods.

Write about Changes

Draw Conclusions What is happening to this plant? Observe the photo and describe the changes that are taking place. Are the changes in the plant changes of state? Tell why or why not.

CHAPTER 2 PHYSICAL SCIENCE EXPERT: PAPER SCIENTIST

People have been making paper for a long time. You might be surprised to find out that papermaking today isn't very much different from what it was thousands of years ago. People have found ways, though, to make the process better. Paper scientist Lucian A. Lucia is one of these people.

Today, most paper is made from wood pulp. Wood pulp is made from ground wood chips that are mixed with chemicals. Wood pulp is used to make the paper you write on at school. Paper towels and paper napkins are also made from wood pulp. Cardboard is made from wood pulp, too. Dr. Lucia is finding better ways to make pulp from wood.

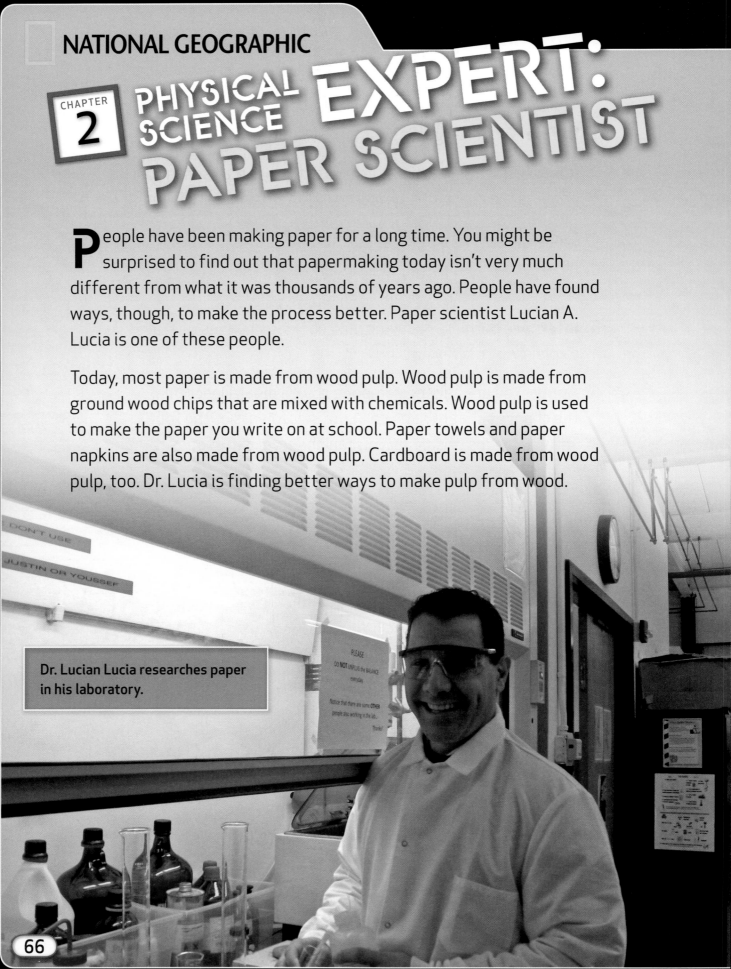

Dr. Lucian Lucia researches paper in his laboratory.

TECHTREK
myNGconnect.com

Student
eEdition

Digital
Library

Dr. Lucia observes the particles that make up wood and pulp. He studies how they are arranged. He looks at the changes that happen when people turn wood into pulp.

Dr. Lucia hopes that his work will lead to a cheaper way of making pulp. He also wants to find a way to make pulp that does not hurt the environment. He enjoys his job. Lucia hopes that his work will improve people's lives.

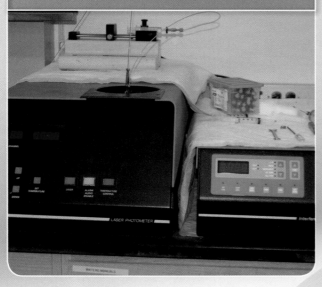

Scientists use special equipment to observe paper and examine its properties.

This very close-up view shows the paper fibers in magazine paper.

BECOME AN EXPERT

Papermaking: A Very Old Tradition

People came up with the idea of paper thousands of years ago. That paper looked different from today's paper, though. The ancient Egyptians made a version of paper beginning around 4000 B.C. They used a plant called papyrus to make sheets to write on. The sheets were also called papyrus. The Egyptians made the sheets by laying wet papyrus fibers close together. Then they laid another layer of fibers across the first layer. The papyrus maker then put a heavy weight on the fibers and let them dry. Once dry, the fibers stuck together to form a sheet.

This carving from Egypt shows a papermaker laying strips of papyrus next to each other. The carving was made between 2446 and 2426 B.C.

TECHTREK
myNGconnect.com

Student
eEdition

Digital
Library

The First Paper

The Chinese invented paper as we know it today. The invention was first reported to the emperor of China in A.D. 105. But scientists now think that paper was being made in China up to 200 years earlier. Early Chinese paper was made from a plant called hemp. The Chinese mixed the hemp with water and beat it to form a pulp. Pulp is a thick mixture of tiny pieces of fiber. This pulp was then put into a mold to dry. When dry, the pulp formed a thin layer—a sheet of paper!—in the bottom of the mold. Later, papermakers used other types of fibers as well.

The Chinese came up with a way to make paper in molds.

Papyrus grew plentifully along the fertile Nile River in Egypt.

Modern Papermaking

Paper is a **solid** material. The process to make paper, however, uses materials in three different **states of matter**! From huge trees to a pulp-like mixture to the sheets of paper that make up this book—how do trees become paper?

Paper companies own or manage forests that provide trees for paper.

All these wood chips will eventually be turned into paper.

states of matter

States of matter are the forms in which a material can exist.

solid

A **solid** is matter that has a definite shape and volume.

How Paper Is Made The first step in making paper is to cut wood into small wood chips. Cutting the wood is a **physical change** . The wood chips are just tiny pieces of wood. Next, the chips are made into a pulp. The pulp is made by grinding the chips or by mixing them with chemicals. Both methods break the chips into tiny pieces. The pulp is gray. But people prefer to have white paper to write on. So, the pulp is bleached. The wood pulp is broken down to substances that are white. The change from wood chips to pulp involves **chemical change** . The pulp will never be wood chips again.

Wood pulp is naturally gray in color.

Bleaching changes the pulp to the white color that you normally see in paper.

physical change

A **physical change** is when matter changes to look different but does not become a new kind of matter.

chemical change

A **chemical change** is a change in which new substances are formed.

Once the pulp is white, it is mixed with a lot of **liquid** water. This mixture ends up being mostly water. So, although the pulp is still a solid, the mixture flows like a liquid. The mixture then passes through big rollers. The rollers start forming the sheet of paper. At the same time, water is pulled out of the mixture.

Mixing pulp with water makes it easy to form the paper into thin sheets.

liquid

A **liquid** is matter that has a definite volume and takes the shape of its container.

To get all of the water out of the paper, the paper goes through a dryer. Inside the dryer, the paper moves between hot rollers. The heat dries the paper. But where does the water go? The water changes to a **gas** —water vapor.

The heated rollers change liquid water in the paper to invisible water vapor.

gas

A **gas** is matter that spreads to fill a space.

After the paper is dry, it is coated with a mixture that makes the paper easier to write on. If the paper is not coated, inks would spread into the paper and writing would not look neat. The paper can also be coated with a clay mixture. The clay mixture makes the paper shiny. Magazine paper is coated with a clay mixture. After the paper is coated, it is dried again. Finally, the paper is cut to size. The trees have changed a lot to become paper! And those changes are permanent.

Large sheets of paper need to be cut to usable sizes.

The process is complete! The paper is formed, dried, and wound onto rolls.

Recycled Paper

Making recycled paper is similar to making regular paper. The biggest difference is that the pulp used in recycled paper is not made from wood. Instead, it is made from used paper. Used paper is mixed with hot water and chemicals. The water and chemicals cause the paper to break down into pulp. They also work to remove ink from the used paper. The pulp is washed to remove the ink and then bleached to make the pulp white. After that, the pulp is made into paper. Making recycled paper is not too different from making paper out of wood pulp!

TECHTREK
myNGconnect.com

Digital Library

Recycling paper reduces the number of trees that have to be cut down to make paper.

Turn and Talk How do papermakers change matter into paper? Form a complete answer to this question together with a partner.

Read Select two pages in this section. Practice reading the pages. Then read them aloud to a partner. Talk about why the pages are interesting.

Write Write a conclusion that tells the important ideas you learned about papermaking. State what you think is the Big Idea of this section. Share what you wrote with a classmate. Compare your conclusions. Did your classmate talk about the different changes in matter that happen during the papermaking process?

Draw Imagine a friend asks you to describe how paper is made. Draw a flowchart to show the process. Label the changes in matter that take place. Then combine your drawing with those of classmates to make a papermaking how-to guide.

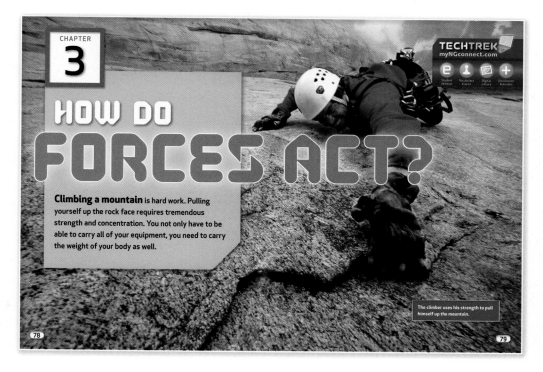

CHAPTER

3

HOW DO FORCES ACT?

Climbing a mountain is hard work. Pulling yourself up the rock face requires tremendous strength and concentration. You not only have to be able to carry all of your equipment, you need to carry the weight of your body as well.

TECHTREK
myNGconnect.com

The climber uses his strength to pull himself up the mountain.

78

79

After reading Chapter 3, you will be able to:

- Identify a force as a push or a pull. **FORCE CHANGES MOTION**

- Describe that if forces are applied to an object that does not move, the forces being applied are balanced. **FORCE CHANGES MOTION**

- Explain that objects move because unbalanced forces are applied to the object. **FORCE CHANGES MOTION**

- Explain that the motion of an object is affected by the size of the force and the object's mass. **FORCE CHANGES MOTION**

- Explain that the speed of an object depends on the time it takes the object to move a certain distance. **SPEED**

- Explain that objects can move in different ways. **KINDS OF MOTION**

- Define friction as a force that slows motion when objects are touching. **FRICTION**

- Identify gravity as a force that pulls objects to the center of Earth. **GRAVITY**

- Science in a Snap! Explain that the speed of an object depends on the time it takes the object to move a certain distance. **SPEED**

HOW DO FORCES

Climbing a mountain is hard work. Pulling yourself up the rock face requires tremendous strength and concentration. You not only have to be able to carry all of your equipment, you need to carry the weight of your body as well.

ACT?

The climber uses his strength to pull himself up the mountain.

SCIENCE VOCABULARY

force (FORS)

A **force** is a push or a pull. (p. 82)

The boy uses a force to try and get the rope away from the dog.

motion (MŌ-shun)

Motion is a change in position. (p. 82)

The ball is in motion.

my
Science Vocabulary

force
(FORS)

friction
(FRIK-shun)

gravity
(GRA-vi-tē)

motion
(MŌ-shun)

speed
(SPĒD)

TECHTREK
myNGconnect.com

Vocabulary
Games

speed (SPĒD)

Speed is the distance an object moves in a period of time. (p. 88)

The runner in front has a greater speed than the other runners.

friction (FRIK-shun)

Friction is a force that acts when two surfaces rub together. (p. 90)

Rough surfaces cause more friction than smooth surfaces.

gravity (GRA-vi-tē)

Earth's **gravity** is a force that pulls things toward the center of Earth. (p. 94)

Gravity makes these balls fall downward.

Force Changes Motion

The child and the dog below are each pulling on a rope. The child and the dog are each using a force to try to get the rope away from one another. A force is a push or pull that can cause objects to be in motion .

The child and the dog are pulling in different directions. However, even though the child and the dog are each pulling very hard, the rope may not move. Why does this happen? The child and the dog and pulling with equal force. When forces are equal, the object the force is being applied to does not move.

The boy and the dog are pulling with equal force, so the rope does not move.

The children below are playing a game of catch. When they throw the baseball to each other, they are applying force to the baseball. The force of each throw puts the baseball in motion. The baseball moves because an unequal force, the throw, has been applied to the baseball.

The mass of the ball affects its motion. Imagine these children trying to play the same game of catch with a bowling ball. The mass of the bowling ball is greater than the mass of the baseball, and so they would have to use more force to throw the bowling ball back and forth. The greater the mass of an object, the greater the force needed to put that object in motion.

The boy uses a pushing force to put the ball in motion.

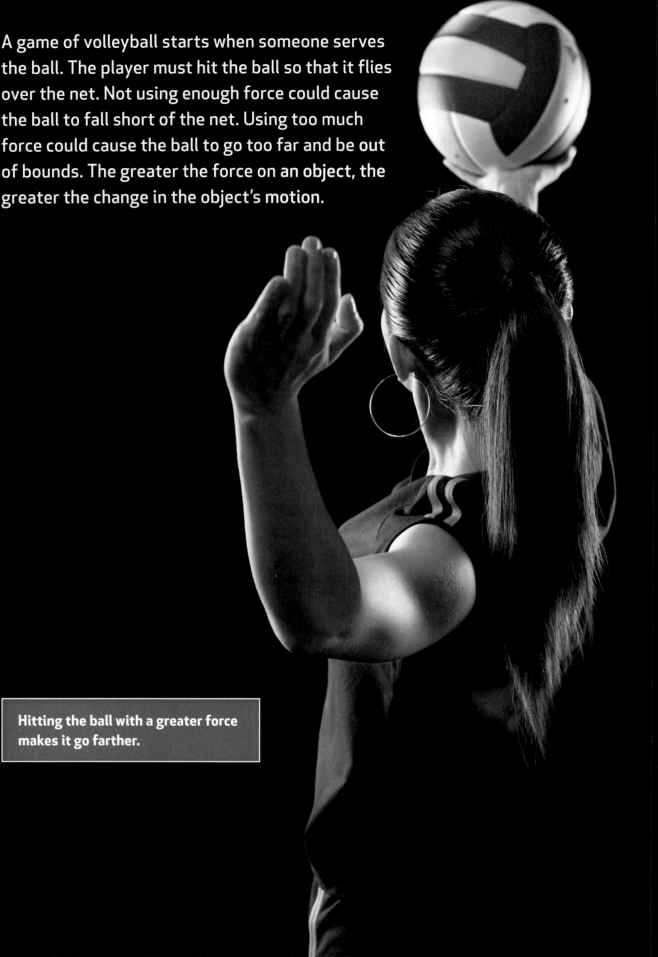

A game of volleyball starts when someone serves the ball. The player must hit the ball so that it flies over the net. Not using enough force could cause the ball to fall short of the net. Using too much force could cause the ball to go too far and be out of bounds. The greater the force on an object, the greater the change in the object's motion.

Hitting the ball with a greater force makes it go farther.

Volleyball players also can change the direction of the ball. They pass the ball to their teammates. They pass the ball in a straight line. The other players use force to change the direction of the ball. The unequal forces applied to the ball makes it change direction and go over the net.

TECHTREK
myNGconnect.com

Digital Library

A player uses less force when setting the ball because she wants the ball to move slowly so a teammate can easily hit it.

A player uses more force when spiking the ball because he wants the ball to move fast so that the other team cannot hit it.

Before You Move On

1. What is force?
2. What causes the motion of an object to change?
3. **Apply** Sometimes a golfer needs to hit a ball a short distance. At other times, a golfer needs to hit a ball a long distance. When does the golfer use more force?

A model train moves along in a straight line on its track. When the train reaches a curve in the track, the direction of its motion may change. It may also slow down as it goes around the curve, and it may no longer go in a straight line. Many objects can move in a variety of different ways, just like the model train.

When the speed or direction of a model train changes, its motion changes.

Sometimes an object's motion follows a pattern. Think about dribbling a basketball. You push the ball down. It hits the ground and bounces back up to your hand. You push it again and it bounces again. The basketball's motion follows a pattern.

Other objects may vibrate. Vibrate means to move back and forth quickly. The tuning fork below is vibrating. Piano tuners use the sound the tuning fork makes when it vibrates to help them tune a piano.

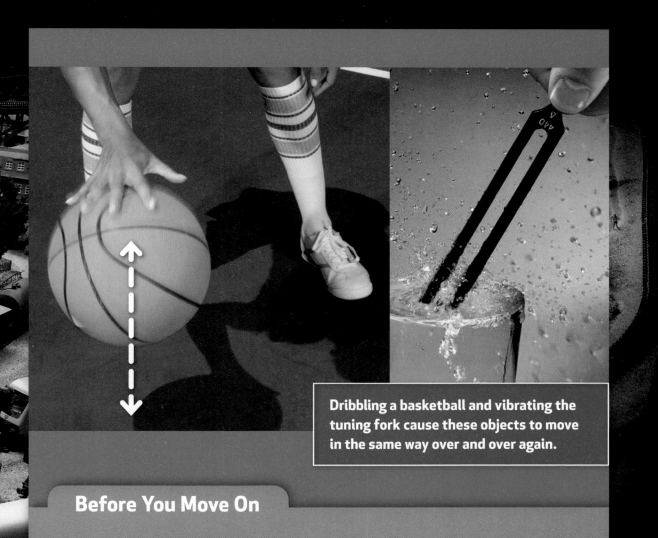

Dribbling a basketball and vibrating the tuning fork cause these objects to move in the same way over and over again.

Before You Move On

1. What are some ways that objects can move?
2. You are bouncing a tennis ball on the ground. Describe its motion.
3. **Apply** A guitar string makes noise when it vibrates. Describe the motion of the guitar string.

Look at the photo below. These children are running in a race. Each of the children is running at a different speed , but they are all going the same distance. Speed is a way to describe the motion of an object. In the race below, a child moving at greater speed changes position faster than a child moving at a slower speed.

The children are all running the race at different speeds.

When the race is over, the speed of each of the children can be found. The children's speed is the distance that they moved in a period of time. Since all of the children ran the same distance, the child that completed the race in the least amount of time had the greatest speed. They were the winner of the race. The child who completed the race in the greatest amount of time had the slowest speed. They finished last.

Before You Move On

1. What is speed?
2. How are speed and changes in an object's position related?
3. Many people are running in a race. How do you know which person runs the fastest?

Friction

Suppose you are running on a gym floor with only socks on your feet. If you tried to stop, you would probably just slide. Stopping on a gym floor with sneakers on is different. You could probably stop right away with sneakers on. The difference has to do with **friction** . Friction is a force that slows motion when surfaces are touching. There is more friction between your sneakers and the floor than there is between your socks and the floor. The friction from your sneakers stops you from sliding.

Science in a Snap! Smooth Move

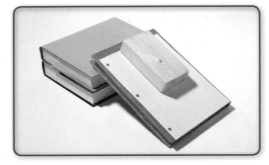

Feel the pieces of notebook paper, sandpaper, and plastic. How are they the same? How are they different?

Make a ramp by leaning a book against two other books. Test each material by placing it on the ramp and letting a block slide down on it.

On which material did the block slide the fastest?

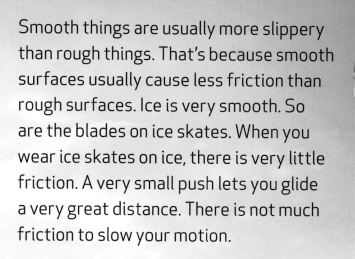

Smooth things are usually more slippery than rough things. That's because smooth surfaces usually cause less friction than rough surfaces. Ice is very smooth. So are the blades on ice skates. When you wear ice skates on ice, there is very little friction. A very small push lets you glide a very great distance. There is not much friction to slow your motion.

TECHTREK
myNGconnect.com

Digital
Library

This ice skater glides easily across the ice because there is very little friction.

The rough tip of an ice skate creates friction. Touching it to the ice slows motion.

How much of a difference does friction really make? You can use a spring scale to compare the force it takes to overcome friction. A spring scale measures force in newtons.

What can you observe in the picture of the spring scale? Based on your observations, how many newtons does it take to drag the basket across the smooth surface of the table?

A spring scale like this one measures force. It takes about 1.5 newtons of force to pull the basket.

It takes about 4.5 newtons of force to drag the same object across the rough carpet. What is the difference? Friction. The smooth surface of the table creates less friction than the rough surface of the carpet. So it takes less force to drag something across the smooth table than it takes to drag something across the rough carpet.

TECHTREK
myNGconnect.com

Enrichment Activities

More force is needed to drag the basket across the carpet than the table.

Before You Move On

1. What is friction?
2. What would it be like to move a piece of furniture across a smooth floor and then across a carpeted floor?
3. **Synthesize** A boy wants to make it easier to grip the handlebars of his bike so his hands don't slip. Should he cover them with a rough material or a smooth material? Explain.

Gravity

In the picture of the juggler, some of the balls are moving up. One of them is moving down. The juggler uses a force to make the balls move up in the air. Another force, called **gravity**, makes them fall down again. Gravity is a force that pulls things toward the center of Earth. Gravity acts on objects without touching them.

Gravity brings the balls back down to the juggler.

Gravity pulls on you, your desk, your school bus, and everything else on Earth. The force of gravity's pull on an object is the object's weight. In other words, the weight of a desk is the force of Earth's gravity on that desk.

Remember that a spring scale measures force. And gravity is a force. So you can measure gravity's force—or weight—with a spring scale. When measured on a spring scale, weight is measured in newtons.

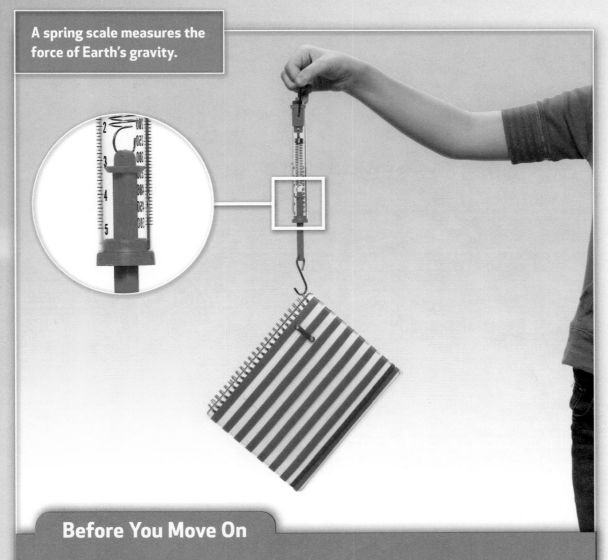

A spring scale measures the force of Earth's gravity.

Before You Move On

1. What is gravity?
2. How are weight and gravity related?
3. **Hypothesize** If there were no gravity, what would happen to a ball if you threw it up in the air?

SUPERCOASTERS

If you like high places and fast speeds, you probably like modern roller coasters. These new roller coasters can travel at rocket-like speeds, climb to great heights, and send riders upside down through loops and corkscrews.

Roller coasters and other rides use forces and motion to thrill their riders.

Using Gravity Many roller coasters use the force of gravity to pull them. A chain pulls the roller coaster car up to the top of a tall hill. When the chain lets go of the car, it rolls down the hill. Gravity pulls the car downward and causes it to speed up. Its high speed keeps it going up the next hill. As the car gets to the top of the hill, it slows down. Once it goes over the next hill, gravity pulls the car downward again.

Coasters of the Future Scientists are working on new ways to make roller coasters faster and even more exciting. One way to speed up roller coasters is to cut down on friction. Roller coasters in the future might use magnets. The cars would float just above the track, without actually touching it. There would be no friction between the cars and the track to slow down the ride.

Many roller coasters get their motion from gravity.

A force is a push or a pull. An object at rest requires a push or a pull to be set in motion. The motion of an object is in the direction of the push or pull. Force changes motion. The greater the force, the greater the change in motion. Friction is a force that slows motion when surfaces are touching. Gravity is a force that pulls objects toward the center of Earth. It acts on objects from a distance.

Big Idea Forces are pushes or pulls that act on objects and cause their motion to change.

FORCES

Push or Pull

Friction

Gravity

Vocabulary Review

Match the following terms with the correct definition.

A. force
B. friction
C. gravity
D. motion
E. speed

1. Force that pulls things toward the center of Earth

2. Force that acts when two surfaces rub together

3. The distance an object moves in a period of time

4. A push or a pull

5. A change in position

Big Idea Review

1. Recall What force acts only when objects are touching?

2. Identify What force puts a roller coaster in motion at the top of a hill?

3. Relate How can the position and motion of an object be related?

4. Compare and Contrast Compare the force of gravity on a car with the force of gravity on a bike. How and why are the forces different?

5. Infer A quickly rolling ball suddenly stops moving. What must have caused the motion of the ball to change?

6. Apply In baseball, a player can hit a home run, which sends the ball a very long distance. Or, a player can bunt, which sends the ball only a few meters. Which action takes more force? Explain.

Write About Forces

Describe What is happening in this picture? What forces are acting on the bucket? Is the bucket in motion? What could change the motion of the bucket?

PHYSICAL SCIENCE EXPERT: ROLLER COASTER DESIGNER

CHAPTER 3

Cynthia Emerick

Roller coasters depend on forces to give riders the thrills they expect. Roller coaster designer Cynthia Emerick takes advantage of those forces as she designs roller coasters.

The Manhattan Express reaches a top speed of 107 kilometers (67 miles) per hour as riders zip past a made-up New York skyline in the middle of Las Vegas.

TECHTREK
myNGconnect.com

Student
eEdition

Digital
Library

NG Science: What is your job?

Cynthia Emerick: My job is to oversee the design, fabrication, and installation of roller coasters and other amusement park rides. I find out what clients want. And then I figure out how to do it in a way that's safe and not too expensive. What it all comes down to is problem solving. My team of engineers and I solve problems every day.

NG Science: What did you do in school to learn how to do your job?

Cynthia Emerick: I always found math and science interesting, so I did well in those classes in high school. In college, I learned about materials and how to put them together to make things. I also learned about what happens when the materials break down—or when the way they are put together breaks down. That's called failure analysis, and that's what helps me understand how to make rides safe.

NG Science: Are there certain projects that you especially liked?

Cynthia Emerick: I am most proud of two projects. I helped design the Manhattan Express at the New York New York Hotel in Las Vegas. It is one of the most widely recognized roller coasters in the world. The other is Revenge of the Mummy— The Ride at Universal Studios. I was in charge of creating the vehicle— the car in which riders sit.

Digital
Library

This ride's special motor launches the cars very quickly to top speed. That power takes the cars up the first hill. From there, gravity and special effects do the rest!

BECOME AN EXPERT

Bobsleds: Using Gravity to Race

If you like racing and cold weather, you might want to try the exciting sport of bobsledding. Bobsledding is a sport in which a team of athletes race down an icy track in a special sled.

Bobsledding got started in the European country of Switzerland in 1877. People took sleds and added steering wheels. They used these bobsleds to slide down icy roads. In 1932, bobsledding became an Olympic sport.

TECHTREK
myNGconnect.com

Digital Library

The team gives the bobsled a running start by pushing it.

 Modern bobsleds are shaped to be as fast as possible.

TECHTREK
myNGconnect.com

Student
eEdition

Digital
Library

Pushing Off

Bobsled teams may have two or four members. A team starts the race at the top of a hill. The race course is like a very long, icy slide. The team does not use a machine to get the bobsled in **motion**. Instead, the team pushes the bobsled. Together, the team members use as much **force** as possible. Once they get the sled moving down the hill, they all jump in.

The push bars fold back into the bobsled once it is in motion.

Once they get going, the team members jump into the bobsled.

motion
Motion is a change in position.

force
A **force** is a push or a pull.

Down They Go

If team members get a good start, the next trick to winning a bobsled race is steering well. One member of the team is the driver. The driver steers by pulling on special handles. They are attached to runners on the bottom of the bobsled. The runners are like the metal parts of ice skates. To go really fast, the driver must keep the bobsled in the best position on the track.

HOW DOES A BOBSLED WORK?

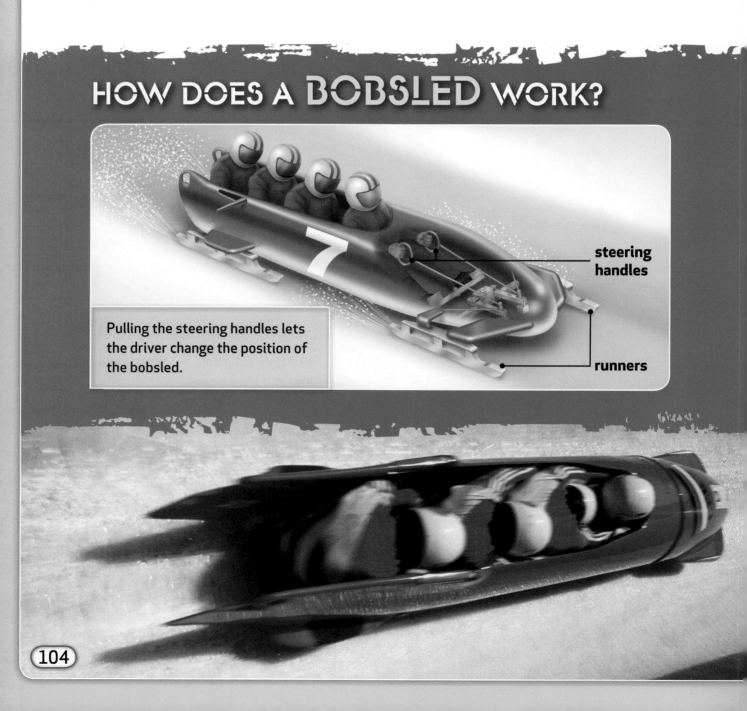

steering handles

runners

Pulling the steering handles lets the driver change the position of the bobsled.

Gravity at Work After the running start of the race, the only force that makes the bobsled keep moving is **gravity**. The heavier a bobsled is, the more gravity pulls on it. But, there is a limit to how much weight you can have in a bobsled. A team of four and its sled cannot weigh more than 630 kilograms (1,389 pounds). If the weight of the four team members and their sled is less than that, they can add weight to make the sled heavier.

Gravity pulls the bobsled down the slope of the course.

gravity

Gravity is a force that pulls things toward the center of Earth.

Smooth and Fast

Ice is slippery, which is why a bobsled can move so quickly on it. To move fast, there must be as little **friction** as possible between the runners of the bobsled and the ice. Athletes keep the runners polished to reduce friction.

However, when team members are pushing the bobsled, they need a lot of friction. To get more friction between their feet and the ice, they wear spiked shoes. The spikes help them grip the ice and push off from it.

There is a lot of friction between the spiked shoes and the ice. The rough surface of the shoes creates friction when the athletes move their feet against the ice.

There is little friction between the bobsled's runner and the ice. The runner slides smoothly on the ice.

friction

Friction is a force that acts when two surfaces rub together.

Top Speeds As with any other race, bobsledding is about moving at the greatest **speed** . At the beginning of the race, the bobsled team starts slowly. But then gravity pulls the bobsled down a course that is almost 1.6 kilometers (1 mile) long. During the race, the bobsled goes faster and faster. By the end of the race, the bobsled team may be going as fast as 128 kilometers (80 miles) per hour!

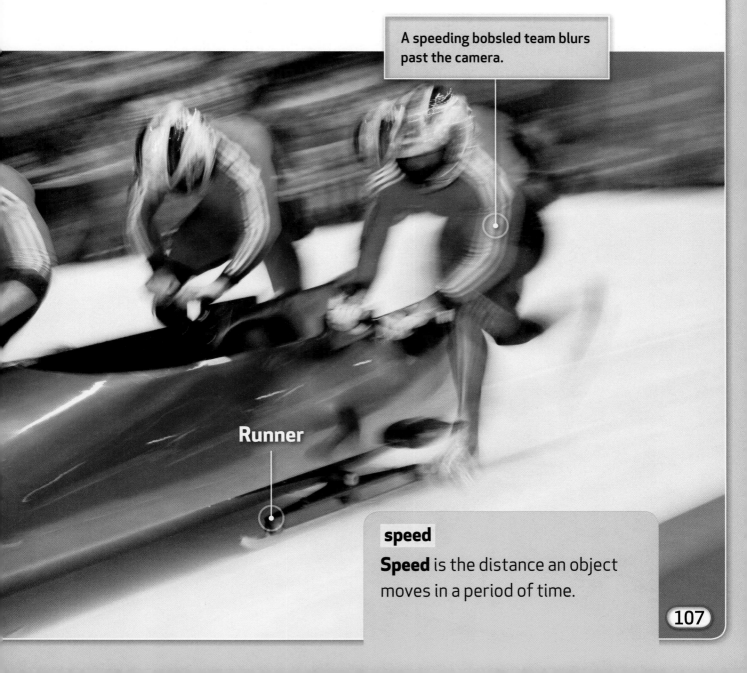

A speeding bobsled team blurs past the camera.

Runner

speed

Speed is the distance an object moves in a period of time.

CHAPTER 3
SHARE AND COMPARE

Turn and Talk How do bobsledders take advantage of different forces to help them move faster? Form a complete answer to this question together with a partner.

Read Select two pages in this section. Practice reading the pages. Then read them aloud to a partner. Talk about why the pages are interesting.

Write Write a conclusion that tells the important ideas about the forces in bobsledding. State what you think is the Big Idea of this section. Share what you wrote with a classmate. Compare your conclusions. Did you recall the effect that friction and gravity have on the speed of the bobsled?

Draw Work with a partner to design a bobsled. Make a detailed drawing. Label the parts of the bobsled. Then present your design to the class, explaining how your bobsled will win races by taking advantage of gravity and reducing friction.

CHAPTER
4

WHAT IS MAGNETISM?

Magnets are fun to play with. They are part of many kinds of toys, such as the shapes and connectors in this photo. Magnets produce a force called magnetism. This force pulls certain parts of the magnets together.

This toy contains magnets. The shapes can be put together to create patterns.

TECHTREK
myNGconnect.com

After reading Chapter 4, you will be able to:

- Describe that magnets can attract and repel other magnets. **MAGNETISM**

- Describe that magnets can attract magnetic materials. **MAGNETISM**

- Classify materials as magnetic and non-magnetic. **MAGNETISM**

- Describe the effects of a magnetic field. **MAGNETIC FIELDS**

- Explain that the force of magnetism decreases as distance increases. **MAGNETIC FIELDS**

- **Science in a Snap!** Describe that magnets can attract magnetic materials. **MAGNETISM**

WHAT IS MAGNE

Magnets are fun to play with. They are part of many kinds of toys, such as the shapes and connectors in this photo. Magnets produce a force called magnetism. This force pulls certain parts of the magnets together.

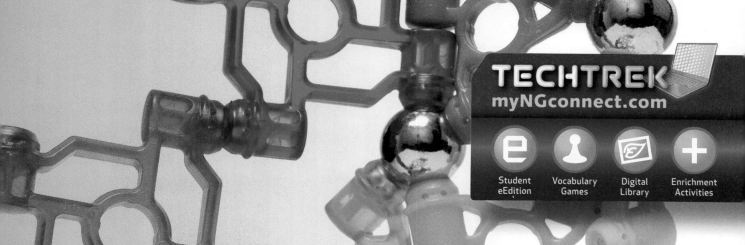

TECHTREK
myNGconnect.com

Student
eEdition

Vocabulary
Games

Digital
Library

Enrichment
Activities

TISM?

This toy contains magnets.
The shapes can be put
together to create patterns.

SCIENCE VOCABULARY

magnet (MAG-net)

A **magnet** is an object able to pull some metals toward itself. (p. 114)

This magnet pulls metal objects that contain iron.

magnetism (MAG-nuh-tism)

Magnetism is a force created by magnets that pulls some metals. (p. 114)

Magnetism holds this magnet to the refrigerator.

pole (PŌL)

The **pole** is the part of the magnet where the force is the strongest. (p. 116)

The forces are the strongest at the poles of a magnet.

my
Science
Vocabulary

attract
(a-TRAKT)

magnet
(MAG-net)

magnetic field
(mag-NET-ik FĒLD)

magnetism
(MAG-nuh-tism)

pole
(PŌL)

repel
(ra-PEL)

attract (a-TRAKT)

To **attract** is to pull toward. (p. 116)

The north pole on a magnet attracts the south pole of another magnet.

S   N   S   N

repel (ra-PEL)

To **repel** is to push away. (p. 116)

The north pole of a magnet repels the north pole of another magnet.

S   N   N   S

magnetic field
(mag-NET-ik FĒLD)

The **magnetic field** is the area around a magnet where there is a pulling force. (p. 120)

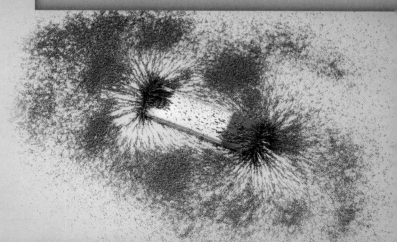

The magnetic field of a magnet is strongest around the poles.

Magnetism

Take a look at the refrigerator door. Do you see the **magnets** ? How do the magnets stay there? A magnet is an object that is able to pull some metals toward it. The magnets stay on the refrigerator because they are pulling on the metal in the refrigerator door. The magnets on this refrigerator are holding up papers. How? They hold up the papers with **magnetism** , a force that pulls some metals. The force is powerful enough that the magnet stays in place even with a piece of paper between it and the metal in the door. Magnetism can act on objects without touching them.

These magnets pull on the metal refrigerator door even through paper.

Metals pulled by magnets are magnetic. Magnets do not pull non-magnetic objects toward them. Look at the photo below. The horseshoe magnet pulls the metal in the paper clip toward it. The paper clips are magnetic.

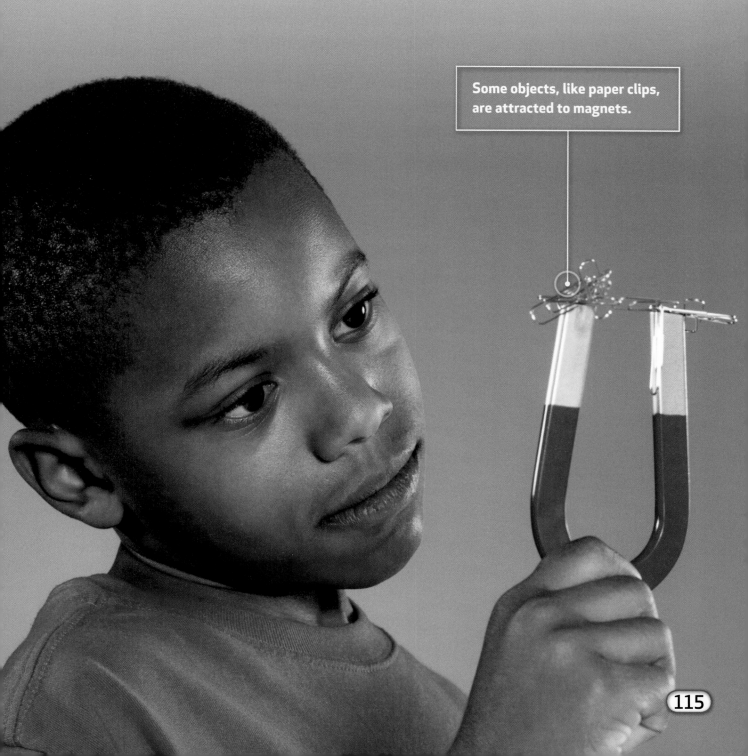

Some objects, like paper clips, are attracted to magnets.

A magnet has two **poles**. These are the parts of the magnet where the force is the strongest. Every magnet has a north pole and a south pole. The north pole of one magnet **attracts**, or pulls toward, the south pole of another magnet. The north pole of one magnet **repels**, or pushes away, the north pole of another magnet. The magnetism at the poles can work without touching other magnets or other materials.

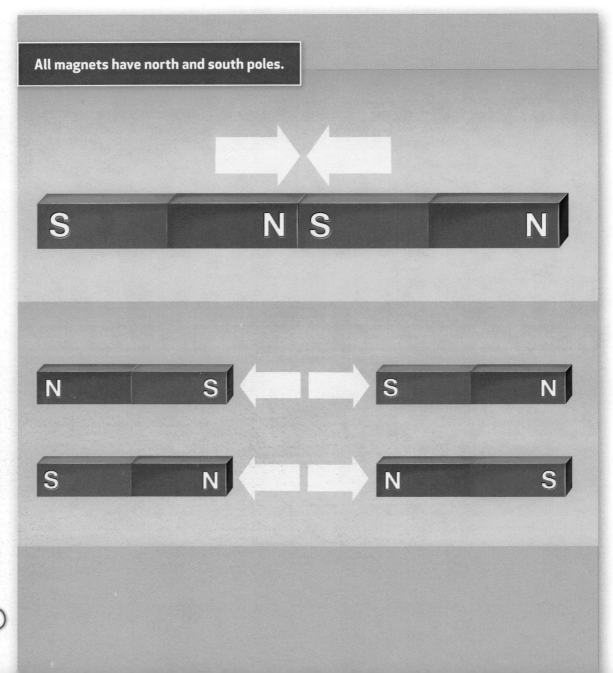

All magnets have north and south poles.

Did you know that the Earth acts as an enormous magnet? It has a north and south magnetic pole just like a regular magnet. A tool called a compass uses Earth's magnetic poles to show directions. There is a magnet in the needle of the compass. The magnet in the needle causes the needle to point toward Earth's magnetic north pole. Compasses help people know the direction they are going.

N

A compass can help you go in the right direction.

Magnets attract materials toward them that are made of iron, nickel, and cobalt. These are all metals.

But a magnet does not attract all metals. A nickel is a coin that does not actually contain the metal nickel. A nickel is not magnetic. Other coins do not have iron, nickel, or cobalt in them. They are not magnetic, either.

Science in a Snap! Make a Magnet

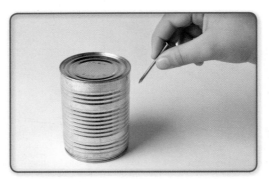

Hold a steel nail close to metal objects. Does it attract the metal objects?

Rub a magnet against the steel nail from the head of the nail to the point. Repeat 20 times. Place the nail near the metal objects.

How did the steel nail change between steps one and two?

Coins and cans are metallic. They are not necessarily magnetic.

You can classify objects as magnetic or non-magnetic based on whether or not a magnet attracts them.

TECHTREK
myNGconnect.com

Enrichment Activities

MAGNETIC OBJECTS

metal tacks

metal button

nail

NON-MAGNETIC OBJECTS

colored pencils

ceramic mug

seashells

Before You Move On

1. Identify a common example of a magnet.
2. What are some non-magnetic objects?
3. **Apply** How would an iron nail and a book behave when a magnet is present?

Magnetic Fields

A **magnetic field** exists around a magnet. It is the area in which there is a pulling force. This pulling force attracts objects to the magnet even though the magnet does not touch the objects.

When an object is close to a magnet, the force of magnetism is stronger. If the object is magnetic, it will be pulled toward the magnet. As a magnetic object moves farther away, the force of magnetism decreases. At a great distance, the object is no longer affected by that magnet.

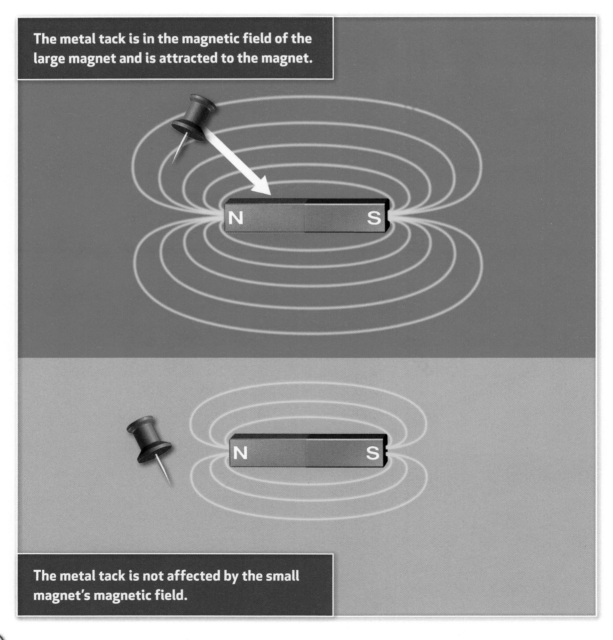

The metal tack is in the magnetic field of the large magnet and is attracted to the magnet.

The metal tack is not affected by the small magnet's magnetic field.

What patterns do you see in the photo? Describe them out loud. Based on your observations, what can you infer about the magnet's magnetic field?

You can learn about the magnetic field of a magnet by looking at patterns. These are iron filings. They are magnetic. You can see how the magnetic field is strongest at the poles of the magnet. That is where the most iron filings can be found. As the distance from the magnet increases, the magnetic field has less effect on the iron filings.

The patterns in the filings at each end of the magnet show where the magnet's strong magnetic field has pulled the filings toward it.

Now look at the patterns in the photo below. How would you describe the pattern? Based on your observations, what can you infer about the magnetic field of this magnet?

This magnet is a horseshoe magnet. The north and south poles are located close to one another. The iron filings appear to be strongly attracted to the end of the magnet with the poles. The iron filings do not seem to be attracted to the curved end of the magnet. This suggests the magnetic field is weaker at the curved end than at the poles.

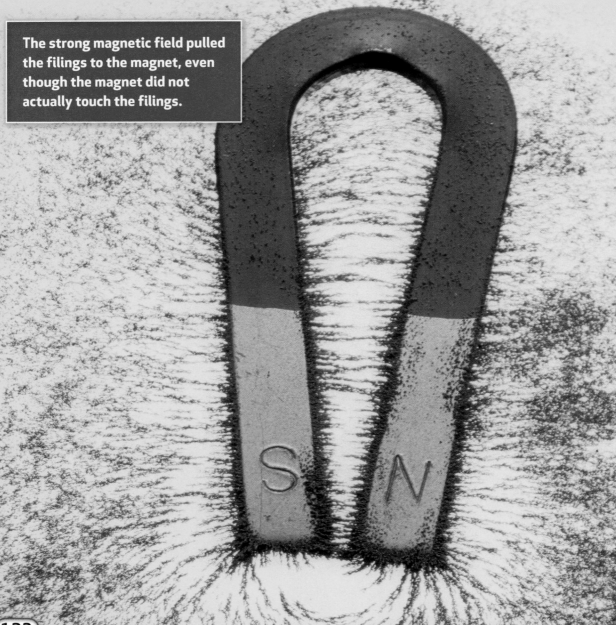

The strong magnetic field pulled the filings to the magnet, even though the magnet did not actually touch the filings.

How many magnets can you see in the photo? Three magnets are red. The other magnet is white. Are these three red magnets stronger than this single white magnet? Yes. You can tell because the three magnets attract more iron filings than the single magnet. This means that the three magnets have a stronger magnetic field than the single magnet.

The magnetic field is not strong enough to pull all of these filings toward the magnets.

Before You Move On

1. What is a magnetic field?
2. Describe what will happen if you hold a magnet near a metal tack.
3. **Infer** What can you tell from the pattern that iron filings make around a magnet?

MAGNETISM AND COMPASSES

As early as 800 B.C., the ancient Greeks knew that a rock, called lodestone, had magnetic properties. Lodestone is a natural magnet. The Greeks knew that it attracted iron. Did the Greeks also notice that a lodestone aligns itself in a north-south direction? We don't know. We do know, though, that early travelers used magnetic compasses as navigation tools. Records show that the Chinese were using magnetic compasses in the 11th century.

Perhaps an early compass was made like this: a thin piece of iron was magnetized by rubbing lodestone on it. The needle was then set on a reed floating in a bowl of water. In time, improvements were made. People figured out how to mount the needle on a pin. Then they could place a card below the needle with the directions written on it—north, south, east, and west.

All compasses use Earth's magnetic field.

This piece of lodestone is a natural magnet. It attracts these iron nails.

Today, magnetic compasses are used in navigation on airplanes, ships, cars, and by people on foot. Hikers and backpackers often carry a compass and map to navigate in remote areas. Because of Earth's magnetic poles, the arrow on the compass always points north. By turning the compass in your hand, you can line up the arrow with the label for north on the face of the compass. This allows you to determine south, east, and west, as well.

A compass is often used with a map to find the location of precise destinations.

A magnet can pull some metals toward itself. Magnetic materials contain iron, nickel, or cobalt. All magnets have two poles where their magnetic field is the strongest. A magnet's north pole attracts the south pole of another magnet. Two like poles repel, or push away from, each other.

Big Idea Magnets produce a force called magnetism that can pull some metals.

Vocabulary Review

Match the following terms with the correct definition.

A. magnetism

B. poles

C. attract

D. repel

E. magnet

F. magnetic field

1. The part of the magnet where the force is the strongest

2. A force created by magnets that pulls some metals

3. To pull toward

4. An object able to pull some metals toward itself

5. To push away

6. The area around a magnet where there is a pulling force

Big Idea Review

1. **Organize** Write the magnetic objects in one list. Write the non-magnetic objects in another list.

coin	eraser	metal tack
paper clip	nail	pencil

Magnetic Materials	Non-Magnetic Materials

2. **Recall** How does a magnet work?

3. **Contrast** How are magnetic and non-magnetic materials different from one another?

4. **Cause and Effect** What happens when you try to push the south poles of two magnets together?

5. **Infer** How would a magnet affect a plastic toy?

6. **Apply** Suppose that you are given two rocks. One of them is lodestone. How can you figure out which one is the lodestone?

Write about Magnets

Explain What does this picture show? Why aren't all of the objects attracted to the magnet?

PHYSICAL SCIENCE EXPERT: MRI RESEARCHER

What does an MRI researcher do? What if a person were sick and doctors weren't sure what was wrong? Wouldn't it be great if they could take a very detailed picture of the inside of the body to figure out what the problem is? Well, they can! And thanks to researcher Vivian Lee, the quality of those pictures is getting better so that doctors are better able to diagnose and cure illnesses and diseases.

These "pictures" are the result of magnetic resonance imaging, or MRI. MRIs help doctors identify or locate many types of conditions.

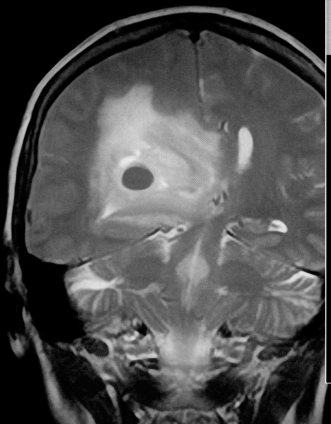

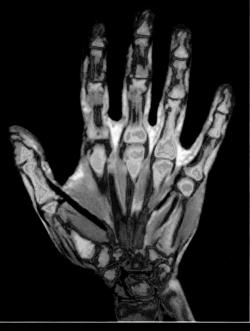

MRIs allow doctors to find problems quickly and safely.

Vivian Lee's MRI research relates to helping people with kidney problems. Other methods of detecting kidney problems are not very good. Using MRI to check kidney function will allow doctors to find problems early and to prevent them from getting worse.

What does all of this have to do with magnets? Inside your body are tiny molecules of water. Within each molecule of water, there is an even tinier particle that acts like a magnet. During an MRI, a very strong magnet is placed next to a person's body. All of those tiny magnets within the body line up. That allows the MRI machine's computer to create an image of the inside of the person's body.

If this all sounds exciting, Lee would agree with you. Researchers' knowledge of MRI technology is advancing quite rapidly. Lee views MRI as a way to make people's lives better. And she finds the opportunity to solve a wide range of problems with MRI technology both interesting and satisfying.

The patient lies very still during the MRI. Movement would create an unclear image, similar to the blurry picture you get if you wiggle while taking a picture with a camera.

BECOME AN EXPERT

Finding the Way Home:
Magnetism and Animals

Have you ever wondered how animals navigate? Although animals do not have a handheld compass, they may have a built-in compass. Because Earth is a large **magnet**, it has **poles**. The poles are where the pull of the magnetic force is strongest. Animals can detect the weaker and stronger parts of Earth's magnetic force. In other words, they use those built-in compasses to find their way.

Loggerhead turtles use the Earth's magnetic force to know which direction to swim.

magnet
A **magnet** is an object able to pull some metals toward itself.

pole
The **pole** is the part of the magnet where the force is the strongest.

TECHTREK
myNGconnect.com

Student
eEdition

Digital
Library

Loggerhead turtles migrate in a loop around the Atlantic Ocean using the forces from Earth's **magnetic field** . In a sense, the magnetic field **attracts** an animal toward the right direction. Or, it **repels** an animal and keeps it from going in the wrong direction. The force of Earth's magnetic field also gives the animals a sense of their location. In addition to using the Earth's magnetic field, the turtles also navigate by landmarks, such as underwater volcanoes or coastlines.

Even baby loggerhead turtles and some other animals that are raised in captivity have this ability. This means using magnetism isn't something that is taught to them by others. They have this ability at birth.

Baby loggerhead turtles have to move quickly from their nest to the ocean to find safety.

magnetic field

The **magnetic field** is the area around a magnet where there is a pulling force.

attract

To **attract** is to pull toward.

repel

To **repel** is to push away.

Many types of birds use the Earth's magnetic field during their migrations. The cells in the heads of many birds contain magnetite. Magnetite is another name for lodestone—a natural magnet. Using **magnetism**, magnetite functions as a compass in the brain by sending directional signals to other parts of the brain. These migrating birds know their precise location based on the strength of Earth's magnetic field.

Cells in the brain of this bird contain a natural magnet that helps the bird navigate.

magnetism
Magnetism is a force created by magnets that pulls some metals.

Thrush nightingales migrate from Sweden to Egypt. They use Earth's magnetic field to navigate. They stop in a certain place in northern Egypt during their migration. Scientists believe that the birds eat extra food in this location so they can survive the long flight over the Sahara.

Thrush nightingales lift off in flight.

Bats also have magnetite in their bodies. To see if bats use magnetism to find their way, scientists studied three groups of bats. One group was exposed to a magnetic field with the directions of the poles reversed from Earth's magnetic field. It made the bats feel that north was south and south was north. The second group was exposed to Earth's natural magnetic field. The third group was not exposed to any magnetic forces.

TECHTREK
myNGconnect.com

Digital Library

Big brown bats, like this one, were used to see if bats use magnetism to find their way.

The scientists took the bats away from their home. The bats that weren't exposed to any magnetic forces flew right home. So did the bats that were exposed to Earth's natural magnetic field. The group that was exposed to the reversed poles flew the opposite direction from their home. Later, they turned around to fly in the right direction. This showed that magnetism affects how bats navigate.

The magnetite in a bat's brain helps it detect Earth's magnetic field.

The mole rat is a blind animal that lives alone in tunnels under the ground. How do you think it navigates? It can't use sight to find landmarks. Scientists believe that the mole rat uses the Earth's magnetic field to navigate.

How do you think scientists could test whether the mole rat uses magnetic fields? As with the bat experiment, scientists planned to expose one group of animals to an artificial magnetic field and keep another group exposed to the natural magnetic field of Earth. The next step for the mole rats, then, would be living and navigating in a maze.

This mole rat does not use sight to navigate. It depends on Earth's magnetic field to find its way.

Scientists found that mole rats that were exposed only to Earth's natural magnetic field put their sleeping nests and food chambers in the southern area of the maze. The mole rats exposed to a magnetic field that was opposite Earth's put their sleeping nests and food chambers in the northern section of the maze. This suggests that mole rats use magnetism to navigate in their dark, underground tunnels.

Mole rats live in a maze of underground tunnels.

Scientists did not expect lobsters to be able to navigate using Earth's magnetic field. Lobsters have simple nervous systems compared to other animals that can do this. Scientists were eager to see whether lobsters use Earth's magnetic field to find their way.

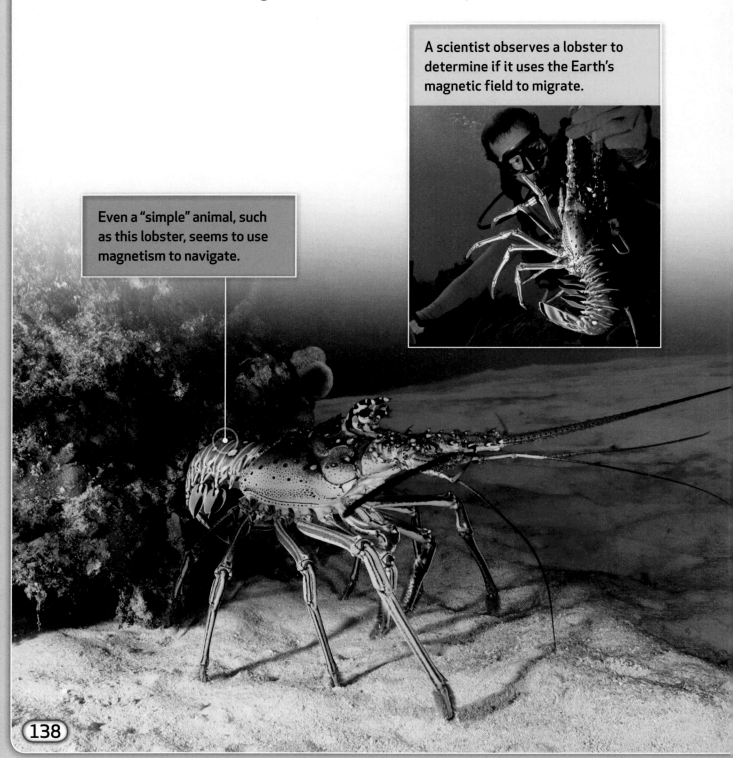

A scientist observes a lobster to determine if it uses the Earth's magnetic field to migrate.

Even a "simple" animal, such as this lobster, seems to use magnetism to navigate.

To test their ideas, scientists captured lobsters. Then, they transported them 12 to 37 kilometers (7 to 23 miles) away from the place where they found them. While they were being moved, the lobsters were in covered containers of seawater. The containers were held by ropes so they swung back and forth. Finally, the lobsters arrived at the test site. Their eyes were covered, so they couldn't see. But when released, they moved in the direction of home. The scientists believe that the lobsters used Earth's magnetic field to find their home.

This lobster is being tagged by a scientist to help the scientist follow the lobster's migration.

CHAPTER
4

SHARE AND COMPARE

Turn and Talk How did the scientists learn about bats' use of magnetism to find their way? Form a complete answer to this question with a partner.

Read Select two pages in this section. Practice reading the pages. Then read them aloud to a partner. Talk about why the pages are interesting.

Write Write a conclusion that tells the important ideas about magnetism and animals. State what you think is the Big Idea of this section. Share what you wrote with a classmate. Compare your conclusions.

Draw Form a group with two other classmates. Have each person draw a picture or diagram of a different animal—the loggerhead turtle, the thrush nightingale, or the Big Brown bat—and how they use the Earth's magnetic poles to find their way. Give each picture a title and include labels. Compare your drawings. Then combine them in a booklet about animals and magnetism.

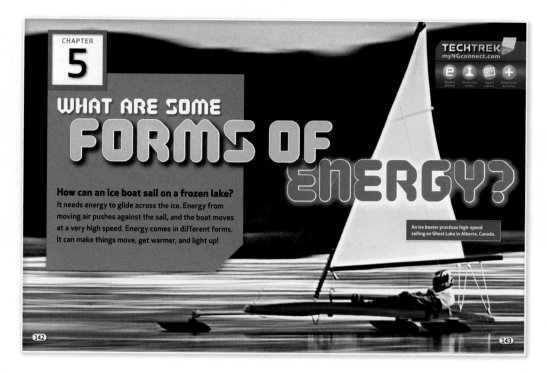

WHAT ARE SOME
FORMS OF
ENERGY?

How can an ice boat sail on a frozen lake?
It needs energy to glide across the ice. Energy from
moving air pushes against the sail, and the boat moves
at a very high speed. Energy comes in different forms.
It can make things move, get warmer, and light up!

TECHTREK
myNGconnect.com

An ice boater practices high-speed
sailing on Ghost Lake in Alberta, Canada.

After reading Chapter 5, you will be able to:

- Describe energy as the source of motion or change. **ENERGY AND WORK**

- Identify and describe many different forms of energy, including mechanical, heat, light, and chemical. **MECHANICAL ENERGY, HEAT, LIGHT, CHEMICAL ENERGY**

- Explain that heat flows between objects until they are all the same temperature. **HEAT**

- Describe heat as the energy produced when substances burn or certain kinds of materials rub against each other. **HEAT**

- Identify common materials that conduct heat well or poorly. **HEAT**

- Explain that light can be reflected from some objects and can pass through other objects. **LIGHT**

- Science in a Snap! Describe heat as the energy produced when substances burn or certain kinds of materials rub against each other. **HEAT**

WHAT ARE SOME FORMS

How can an ice boat sail on a frozen lake?

It needs energy to glide across the ice. Energy from moving air pushes against the sail, and the boat moves at a very high speed. Energy comes in different forms. It can make things move, get warmer, and light up!

TECHTREK
myNGconnect.com

Student
eEdition

Vocabulary
Games

Digital
Library

Enrichment
Activities

OF ENERGY?

An ice boater practices high-speed sailing on Ghost Lake in Alberta, Canada.

SCIENCE VOCABULARY

energy (EN-ur-jē)

Energy is the ability to do work or cause a change. (p. 146)

The players use energy to play the game.

mechanical energy (mi-CAN-i-kul E-nur-jē)

The **mechanical energy** of an object is its stored energy plus its energy of motion. (p. 148)

The mechanical energy of the marble is its stored energy plus its energy of motion.

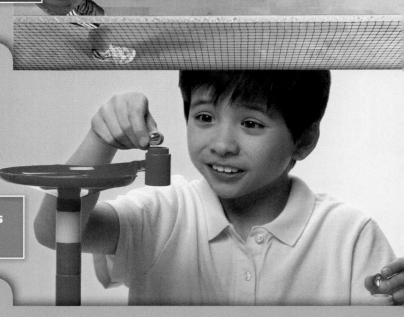

heat (HĒT)

Heat is the flow of energy from a warmer object to a cooler object. (p. 150)

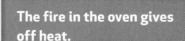
The fire in the oven gives off heat.

my Science Vocabulary

chemical energy
(KEM-i-kul EN-ur-jē)

light
(LĪT)

energy
(EN-ur-jē)

mechanical energy
(mi-CAN-i-kul E-nur-jē)

heat
(HĒT)

TECHTREK
myNGconnect.com

Vocabulary
Games

light (LĪT)

Light is energy that can be seen. (p. 156)

Light reflects off of the cave wall.

chemical energy
(KEM-i-kul EN-ur-jē)

Chemical energy is energy that is stored in substances. (p. 158)

There is chemical energy in the food that the girls are eating.

Energy and Work

Whack! Players hit the birdie with their rackets. The birdie changes direction and goes flying over the net. Playing a game of badminton is a fun activity, but scientists would say the players are actually doing work. Work is using a force to make something move or change. In order to do work, you must have **energy** .

The player uses energy in a game of badminton.

You use energy every time you move. You transfer energy when you move something else. This player is transferring energy from her arm to the racket. The racket transfers energy to the birdie and puts it in motion. All objects can have energy and do work.

The player uses a racket to change the direction of the birdie.

Before You Move On

1. What is energy?
2. How are work and force related?
3. **Apply** Give an example of energy transferring from one object to another.

Mechanical Energy

These marbles are not moving, but they still have energy. They have stored energy. The higher marble has more stored energy than the lower marble because of its position. It has a longer way to roll to the bottom of the toy. When the marbles are in motion, they have energy of motion. The marble's **mechanical energy** is stored energy plus energy of motion.

As the marbles start to roll, their stored energy turns into energy of motion. All moving objects have the energy of motion. The marble's stored energy continues to turn into energy of motion until the marble gets to the bottom of the toy. When the marble stops, it no longer has energy of motion.

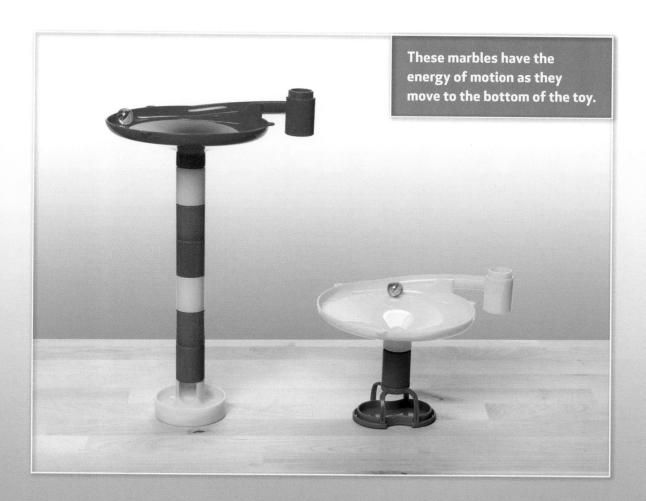

These marbles have the energy of motion as they move to the bottom of the toy.

Before You Move On

1. What is mechanical energy?
2. How is stored energy changed to energy of motion?
3. **Apply** Does a skier standing at the top of a mountain have more or less stored energy than a person standing at the bottom of the mountain?

Heat

Think about a pizza in an oven. Heat from the oven spreads through the pizza. The crust of the pizza starts to cook. Sauce bubbles and cheese melts. Finally the entire pizza is hot and ready to eat.

How did the pizza get hot? Heat from the pizza oven flowed to the cooler pizza. Heat raised the temperature of the cooler pizza. Heat is the flow of energy from a warmer object to a cooler object. Heat raises the temperature of objects.

A brick oven cooks a pizza with heat from a fire.

Look again at the pizza. The pizza outside the oven is not the same temperature as it was. How did the pizza cool off? The heat energy that was in the pizza is greater than the heat in the air around it. The heat moves from the warm pizza to the cooler air. Soon, the heat energy in the pizza will be the same as the heat energy in the air. The temperatures of the pizza and surrounding air will be the same.

Science in a Snap! Getting Warmer

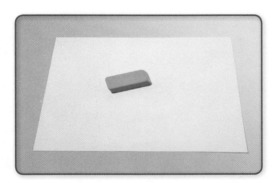

Get an eraser and a piece of paper. Touch both items to feel their temperature.

Rub the eraser quickly on the paper for 30 seconds. Touch both items again.

What happened to the temperature of the eraser and the paper?

Heat Causes Motion Heat energy can be found in many places, from the fire in a pizza oven to the heat the sun provides.

Heat causes motion, even though you may not be able to see it. When the temperature of an object rises, the particles that make up the object begin to move. Those particles bump into other particles, making them move too. Anything that can make particles move faster can heat an object.

This metal has a lot of heat energy. The particles that make up the metal are moving very quickly.

You already know that adding heat energy to an object, such as the pizza in the oven, can cause the object to get warmer. However, fire is not the only way to get the particles in an object to move. For example, imagine that you are standing outside on a chilly day. You have no gloves or pockets. What can you do? You can rub your hands together! This makes the particles in your skin move. This puts heat energy right into your hands. The heat energy that you added to your hands by rubbing them together made your hands warmer.

On a cold day, you can rub your hands together to make them warm. Try it for yourself!

You might have noticed that some objects heat quickly, while others heat very slowly. Good conductors of heat allow the energy to flow through them easily. Metals such as silver, copper, aluminum, and iron are good conductors of heat.

GOOD CONDUCTORS OF HEAT

Enrichment Activities

Iron conducts heat very well. Iron is used to make tools for cooking. Many pots and pans are made of metals including iron.

The silver in the car window conducts heat and defrosts the window.

Steel is another good conductor of heat.

Cloth and rubber objects do not conduct heat well. We can use these materials to protect us from hot things such as a pot on a stove. Wood, glass, and plastic are other materials that do not conduct heat well.

POOR CONDUCTORS OF HEAT

Pot holders are made out of cloth. They protect your hands when you pull something out of the oven.

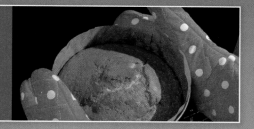

Plastic, cloth, and leather conduct heat poorly. These materials make great safety clothing when you are working with heat like these firefighters.

Before You Move On

1. What is heat?
2. Does a wooden spoon conduct heat well or poorly? Explain your answer.
3. **Apply** Explain how your toast is heated in a toaster.

Light

How many things can you think of that give off light? Lamps, the sun, fireworks, and even lightning bugs give off light. Light is energy that you can see. The scientist in this photo is looking at ancient carvings on a cave wall. He couldn't do this without light from his flashlight. Light travels in a straight line until it hits an object. The light from the flashlight hits the cave wall. The cave wall reflects the light. This means the light bounces off the cave wall. The light enters the eyes of the man, and he is able to see the wall.

This scientist uses light to help him see ancient carvings on a cave wall.

We need light to see things. Sometimes things look different because of the way light hits them. Look at the brush in this glass of water. It looks bent. This is because light can bend when it goes through a new material. When we take the brush out of the water, it looks straight again.

Light bends when it goes through some materials.

Before You Move On

1. What is light?
2. How does light allow you to see objects?
3. **Infer** How does light change as it passes through water?

157

Chemical Energy

From where do you get your energy? Like many other living things, you get your energy from food. Once the food is in your body, your body uses the **chemical energy** stored in the food.

You use the chemical energy in food to keep your body moving and growing. Most of the chemical energy is changed into energy of motion. Your heart beats, your lungs move as you breathe, and your blood moves through your body. You can run and stretch. Some of the chemical energy in your body is even changed into heat energy that helps keep your body warm.

The chemical energy in the foods the girls are eating helps their bodies move and grow.

The chemical energy in food is only one type of chemical energy. You probably know many more. Striking a match releases chemical energy. When the chemical energy stored in the tip of the match is rubbed against a rough surface, it changes very quickly to heat energy. That is why a match catches fire.

Gasoline that is used in cars has stored chemical energy. When gasoline is burned, it changes the chemical energy into heat and energy of motion, making the car move. When all of the gasoline is burned, the gas tank is empty, and the car will not move.

The chemical energy in the gasoline can change to the energy of motion to get you from place to place.

Before You Move On

1. What is chemical energy?
2. How are food and gasoline alike and different?
3. **Apply** Give an example of how chemical energy can be changed to heat energy.

SOLAR POWER
ENERGY FROM THE SUN

The sun provides us with more energy than we could ever use. It provides light and gives plants the energy to make food. There is still plenty of energy to spare. How can we start using more of the sun's energy?

Solar cells collect light and change it to electrical energy when the sun is shining.

The sun gives light. People can collect the light energy from the sun. Let's say that many mirrors reflect the sun's light onto one spot. The heat created at that spot would be very high. Scientists use this heat to warm water. The water boils and changes to steam. The steam can then cause the parts inside a motor to start moving. This motion can generate electrical energy.

Using sunlight and steam to make electrical energy protects the environment. No gas or coal is burned. This means that generating the electrical energy does not cause pollution. And the sunlight is free!

Solar power is a good source of energy, but we still have trouble storing the energy. Scientists are working on ways to collect a larger amount of energy and develop the ability to store it for the future.

The light from the sun bounces off the mirrors and heats water to make steam. The steam generates electrical energy.

Conclusion

Energy is the ability to do work or cause a change. It comes in many forms such as mechanical, light, and chemical. Energy can change from one of these forms to another form. Heat is the flow of energy from a warmer object to a cooler one. All of these forms have the ability to do work or cause a change.

Big Idea Energy, including mechanical, heat, light, and chemical, is the ability to do work or cause a change.

Mechanical Energy

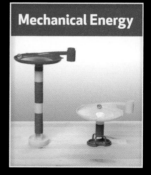

Heat Energy

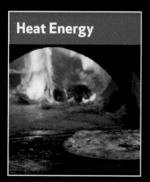

Light Energy

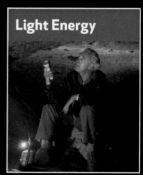

Chemical Energy

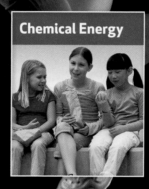

Vocabulary Review

Match the following terms with the correct definition.

A. energy

B. light

C. heat

D. chemical energy

E. mechanical energy

1. The flow of energy from a warmer object to a cooler object

2. The ability to do work or cause a change

3. Energy that is stored in substances

4. A form of energy that can be seen

5. An object's stored energy plus its energy of motion

Big Idea Review

1. **Recall** What is energy?

2. **Restate** Use your own words to define mechanical energy.

3. **Compare and Contrast** Write one sentence explaining how light energy and heat energy are the same, and one sentence explaining how light energy and heat energy are different.

4. **Explain** Explain how you could use heat energy.

5. **Draw Conclusions** What happens if water coming out of a hose runs out of energy of motion?

6. **Describe** Give an example of how chemical energy can cause motion.

Write About Energy

Explain What kind of energy is this girl using?

CHAPTER 5 · PHYSICAL SCIENCE EXPERT: ENGINEER

Allison Gray is an engineer. She is part of the National Renewable Energy Laboratory's concentrating solar power research team. Allison first became interested in solar energy while she was a college student. "There was a renewable energy research center on campus," she recalls. "Working there was my first opportunity to get involved in the industry. Once I had started working in this area, I knew this was the field I wanted to work in."

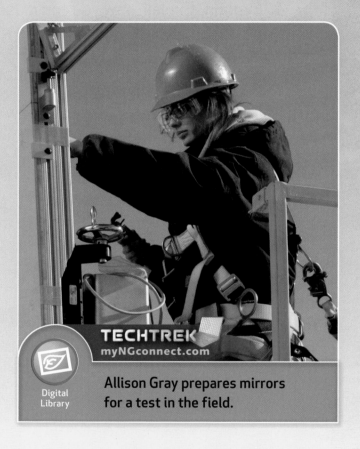

TECHTREK
myNGconnect.com

Digital Library

Allison Gray prepares mirrors for a test in the field.

Allison's research focuses on concentrating solar thermal power. This technology uses mirrors to reflect and concentrate sunlight onto receivers. These receivers collect solar energy and convert it into heat. Heat energy can be used to produce electrical energy.

TECHTREK
myNGconnect.com

Student
eEdition

Digital
Library

"The best part of my job," Allison says, "is that I get to work with many different companies and help them characterize their solar collectors." Depending on the company and their needs, Allison can provide performance information or help to make design changes.

Solar energy is a passion for Allison. "All renewable energy technologies are important," she says. "They should be the main sources for our energy. Solar energy is one renewable technology that should be used. Over time, nonrenewable resources will decrease. Renewable sources will still be abundant." She pauses for a moment, thinking. She says, "Doesn't it make sense to depend on infinite resources?"

These solar panels collect the sun's energy and help transform it to electrical energy.

BECOME AN EXPERT

The Grand Coulee Dam:
When Mixing Energy and Water Works

Scientists are always looking for better ways to get **energy** . One way is using moving water. The Grand Coulee Dam in Washington State was built in 1942 to collect the energy from the water that moves through it. People can then use this energy to power the things they need.

The Grand Coulee Dam changes the movement of water into energy that people can use.

energy
Energy is the ability to do work or cause a change.

TECHTREK
myNGconnect.com

Student
eEdition

Digital
Library

Using moving water to create energy is not a new idea. For many centuries, farmers have used moving water to grind grain. A large waterwheel was put in motion by flowing water. The wheel turned a stone. The stone turned and crushed the grain into flour. Even today, you will see mills next to fast-moving streams. The flowing water causes the stone to do work. The Grand Coulee Dam collects energy in a similar way.

People as far back as the Roman Empire used waterwheels, like this one in Tennessee, to grind grain into flour.

Today we have modern tools to get energy from moving water. We use hydroelectric plants like the ones at the Grand Coulee Dam to make electrical energy.

The Grand Coulee Dam blocks a river. Gates in the dam let some water flow through. Inside the gate, the water moves quickly past a large fan. The force of the water spins the fan. The **mechanical energy** of the water turns into electrical energy. This electrical energy flows through wires to homes and businesses. The diagram below shows more about how a hydroelectric plant works.

HYDROELECTRIC PLANT

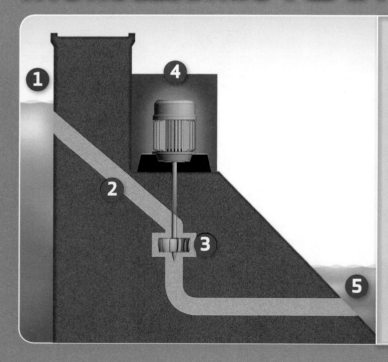

1. Water from the lake enters the hydroelectric plant.

2. Passages that can be opened and closed force water through the plant at high speed.

3. The force of the water causes large fans to move.

4. Generators attached to the fans change the mechanical energy from the moving water to electrical energy.

5. The water continues down the river.

mechanical energy

The **mechanical energy** of an object is its stored energy plus its energy of motion.

The Grand Coulee Dam is so large that it actually has four power plants inside.

The top of the dam is almost 1.6 kilometers (1 mile) across. It is one of the largest concrete structures in the world. The Grand Coulee Dam has enough concrete to build a highway stretching from Seattle to Miami!

The Grand Coulee Dam not only provides power, it also protects the land nearby from flooding.

Hydroelectric power, like the power the Grand Coulee Dam provides, brings **light** and **heat** to millions of homes and businesses. In addition, the water is a clean renewable resource. However, building a hydroelectric dam requires careful planning. The dam should not cause damage to the environment. It also has to be very strong. The dam has to hold back millions of gallons of water!

light
Light is energy that can be seen.

heat
Heat is the flow of energy from a warmer object to a cooler object.

People have been using energy from moving water throughout history. Scientists continue to look for ways to get the energy we need. Oil, coal, and other fossil fuels that use stored **chemical energy** often give off pollution that can hurt the environment. Clean sources of power such as moving water can help keep our planet healthy in the future.

The electrical energy used by people in this part of Washington comes from the energy of motion in the flowing river waters.

chemical energy
Chemical energy is energy that is stored in substances.

171

CHAPTER
5

SHARE AND COMPARE

Turn and Talk) How can the motion of water be used to create energy that people can use? Form a complete answer to this question together with a partner.

Read) Select two pages in this section. Practice reading the pages. Then read them aloud to a partner. Talk about why the pages are interesting.

 **Write**) Write a conclusion that tells the important ideas about hydroelectric power. State what you think is the Big Idea of this section. Share what you wrote with a classmate. Compare your conclusions.

 **Draw**) Form groups of four. Have each person draw one part of a hydroelectric plant. Label each part. Then put the drawings together to show how water moves through a hydroelectric plant to create electrical energy.

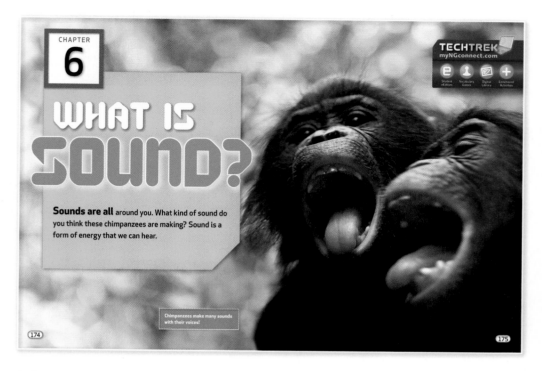

CHAPTER

6

WHAT IS SOUND?

Sounds are all around you. What kind of sound do you think these chimpanzees are making? Sound is a form of energy that we can hear.

Chimpanzees make many sounds with their voices!

174

175

After reading Chapter 6, you will be able to:

- Recognize that vibrating objects make sound, and sound can make things vibrate. **SOUND TRAVELS, VOLUME**

- Understand the relationship between volume and vibrations. **VOLUME**

- Understand that pitch is caused by vibrations. The faster the vibrations, the higher the pitch. **PITCH**

- Science in a Snap! Recognize that vibrating objects make sound, and sound can make things vibrate. **VOLUME**

WHAT IS SOUND?

Sounds are all around you. What kind of sound do you think these chimpanzees are making? Sound is a form of energy that we can hear.

Chimpanzees make many sounds with their voices!

TECHTREK
myNGconnect.com

Student
eEdition

Vocabulary
Games

Digital
Library

Enrichment
Activities

SCIENCE VOCABULARY

sound (SOWND)

Sound is a form of energy that you hear. (p. 178)

These cars make sound that the drivers can hear.

vibration (vī-BRĀ-shun)

A **vibration** is a rapid, back-and-forth movement. (p. 178)

All sound starts as a vibration. When this glass shattered, the vibrations of the breaking glass created sound.

my
Science
Vocabulary

pitch
(PICH)

sound
(SOWND)

vibration
(vī-BRĀ-shun)

volume
(VOL-yum)

TECHTREK
myNGconnect.com

Vocabulary
Games

volume (VOL-yum)

Volume is the level of loudness of a sound. (p. 182)

These large bells make a sound with a loud volume.

pitch (PICH)

Pitch is how high or low a sound is. (p. 186)

Instruments of different sizes make sounds of different pitches.

Sound Travels

Look at the pictures on this page. What **sounds** do you think these objects make? Sound is a form of energy that you can hear. Sound starts as a vibration. A **vibration** is a rapid, back-and-forth movement. Think of how a rubber band acts when you stretch it out and pluck it. When the rubber band vibrates, it pushes against tiny particles in the air. These particles push against other particles. The particles vibrate, and the vibrations make invisible waves of sound that travel through the air. You can see the vibrating rubber band, but you can't see the waves going through the air. As the waves move through the air and to your ear, you hear sounds.

Vibrations from these and many other sources create the sounds that you hear.

There are many sources of sound all around you. Lightly touch your throat as you sing or talk. Can you feel your vocal cords vibrating? Those vibrations make sound that comes out as your voice. If you drop something, it makes vibrations when it hits the floor. You hear the vibrations as a bang or a crash. Some machines use energy to create vibrations. A doorbell's vibrations sound like a bell or buzzer.

Look at the tree in the photo. When it broke in half, it made a sound. But how exactly did the waves travel?

When the tree snapped, the materials in the tree vibrated. Then those vibrations created waves that traveled through the air. If people were in the forest, they could hear the sound of the breaking tree when the waves traveled through the air to their ears. Many animals can hear sounds, too, just as people can.

The vibrations that make sound can travel through other materials. They can travel through water. Water, like air, has tiny particles. Vibrations push those tiny particles in the water. The moving particles create waves of sound that move through the water.

When the tree broke, the vibrations traveled through the air to make sound.

Sound can travel through many solid materials, too. Think about being in your bedroom with the door closed. Someone calls you to dinner. The sound traveled through the air, through your door, and then through the air again to get to your ear. Tiny particles in the door vibrated to form waves that kept the sound moving.

Sound can travel through water. Humpback whales can hear sounds from miles away.

Before You Move On

1. What creates sound?
2. Name a sound you can hear in your classroom. What vibrates to create that sound?
3. **Apply** How do you produce sound when you knock on a door?

Volume

Sounds can be loud or soft. **Volume** is the level of loudness of a sound. The vibrations determine the volume of a sound. Larger vibrations cause louder sounds. Smaller vibrations cause softer sounds.

Digital Library

A **LOUD** SOUND

Bells like these produce large vibrations, making very loud sounds.

A **SOFT** SOUND

These small bells produce very small vibrations. The sounds you can hear are soft.

Sounds at very loud volumes can damage a person's hearing. Tiny hair cells inside the ear change waves into electrical signals that travel to the brain. If sounds are too loud, or if loud sounds last a long time, they can damage these hair cells. It is important to protect your hearing. If you use earphones, you should keep the volume of your music low. People who work in loud places, such as in factories or at construction sites, should wear protective ear coverings.

Airplane engines are very loud. Airport workers protect their ears when working near the planes.

The powerful engines of race cars make big vibrations. The volume of a passing race car is very loud!

Loud and Soft Sounds   Look at the pictures below. Which picture shows soft sounds? Which picture shows loud sounds? We can change the volume of our voices by making our vocal cords vibrate more or less. We can make louder sounds by pushing more air through our vocal cords. That makes them vibrate more, so the noise is louder. Musicians play more forcefully when they want the music to be louder. If you turn a faucet on low, the gentle vibration makes a soft sound. If you turn up the water, the rushing water vibrates more, and the sound is much louder.

A quiet conversation

Shouting

Science in a Snap! Making Vibrations

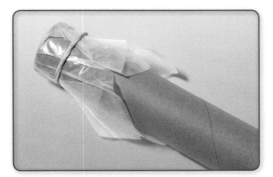

Cut a square of waxed paper and place it over one end of a cardboard tube. Secure the waxed paper with a rubber band.

Speak or hum into the open end of the tube.

What happens to the waxed paper?

A cheering crowd

Before You Move On

1. What is the relationship between vibrations and volume?
2. Compare the volume of sound that occurs when you set your backpack down gently with when you drop your backpack. Explain the difference.
3. **Infer** If you hit a pan lid and a garbage can lid, which would make a louder sound. Why?

Pitch

Did you know that every person on Earth has a unique voice? Like your fingerprints, your voice can identify you. One characteristic of your voice is its pitch. **Pitch** is how high or how low a sound is. Pitch varies by the speed of the vibration. The faster the vibration, the higher the pitch. The slower the vibration, the lower the pitch.

Vibrations of your vocal cords determine the pitch of your voice. The faster your vocal cords vibrate, the higher the pitch of your voice.

These children have high-pitched voices. As they grow, their vocal cords will change. Their voices will change pitch.

Likewise, the slower they vibrate, the lower the pitch is. The speed at which your vocal cords vibrate depends on the length and thickness of the cords. Women's vocal cords are shorter and vibrate at a faster rate than men's. This is why women tend to have higher-pitched voices than men. As children grow, their vocal cords thicken, causing a change in the pitches of their voices.

High Pitch and Low Pitch You hear sounds of many different pitches every day. A truck horn, for example, is a sound with a very low pitch. A whistle has a high pitch. Musical instruments also make sounds of different pitches. Look at the photo on this page. The largest instrument's strings vibrate slowly, making a low-pitched sound. Which instrument do you think has the highest pitch?

While the mariachi instrument in the center makes a low-pitched sound, the instrument on the right would make a sound with a higher pitch.

A violin and a bass are both members of the string family of instruments. They have very different sounds, however. The shorter, finer strings of the violin allow for faster vibrations. The longer, thicker strings of the bass vibrate at a slower rate. This makes the pitch of the violin higher than the pitch of the bass. A viola is smaller than a bass but larger than a violin. Would the pitch of a viola be higher or lower than the pitch of a bass? The strings are finer and shorter, allowing them to vibrate at a faster rate. This vibration gives a viola a higher-pitched sound than the bass.

TECHTREK
myNGconnect.com

Enrichment
Activities

Both instruments can make sounds of different pitches, but the violin's pitches are generally higher and the bass's pitches are generally lower.

Bass

Violin

Before You Move On

1. What is the relationship between vibration and pitch?
2. Classify these sound sources as generally high-pitched or low-pitched: children's voices, racecars, men's voices.
3. **Apply** Jonas made a guitar by wrapping rubber bands around a shoebox. What could Jonas do to make each "string" of his guitar have a different pitch?

NATIONAL GEOGRAPHIC

BREAKING THE SOUND BARRIER

You know that sound travels, but how fast does it go? The speed of sound is affected by the medium through which it is traveling and by the temperature. The speed of sound in the air is approximately 331 meters (1086 feet) per second. The speed changes in colder and hotter air. Sound travels faster in warmer air. It travels more slowly in cooler air. To "break the sound barrier" means to travel faster than the speed of sound. Chuck Yeager was the first man to break the sound barrier in an aircraft in 1947. In 1953, Jackie Cochran became the first woman to break the sound barrier.

Chuck Yeager

Jackie Cochran

A military jet breaks the sound barrier.

Did you ever see a jet high in the sky in one place while the sound it was making seemed to come from another place? Did it seem as if it were outrunning its own sound? That's because the jet was flying at supersonic speed, or faster than the speed of sound. As a jet travels through the air, it creates waves, much like water waves that a boat creates. As the jet flies faster than the speed of sound, the waves can't move fast enough and bump into each other. The result is a thunder-like sound called a sonic boom. Look at the photo on this page. This jet is traveling at supersonic speed. As a jet breaks the sound barrier, a collar of water vapor forms around its body. Scientists call this a vapor cone.

Vapor cone

This jet is traveling faster than the speed of sound, which is approximately 1,225 kilometers per hour.

Conclusion

Sound is a form of energy that you hear. Vibrations make sound waves that travel from the source, through a medium, and to a receiver. One property of sound is pitch. Fast vibrations make sounds with high pitches. Slow vibrations make sounds with low pitches. Another property of sound is volume, or how loud or soft the sound is.

Big Idea Sound is a form of energy that you hear.

Wolves have vocal cords. They can make both high-pitched and low-pitched sounds.

Vocabulary Review

Match the following terms with the correct definition.

A. sound   **1.** A rapid, back-and-forth movement

B. volume   **2.** A form of energy that you hear

C. pitch   **3.** How high or low a sound is

D. vibration   **4.** The level of loudness of a sound

Big Idea Review

1. List Think of the sounds you hear on a school day. Tell what vibrates to make those sounds.

2. Restate Tell how a high-pitched sound differs from a low-pitched sound.

3. Sequence Tell what happens from the moment you stomp your foot until someone hears the sound.

4. Compare and Contrast Compare and contrast the voice of a child to the voice of an adult. What is the same about the sounds? What is different about them?

5. Analyze Listen to the sounds around you. Then focus on one of them. Write a description of the sound, including its volume and pitch. Read the description to a classmate. Can he or she identify the sound based on your description?

6. Infer A guitar player is tuning a guitar. One string is making a pitch that is too low. What does the guitar player do to the string so that it makes a higher pitch?

my SCIENCE notebook

Write About Sound

Look at the photograph. What can you conclude about the sounds each instrument will make? What causes the sound? Which instrument will have a sound with a higher pitch? How might their volumes differ?

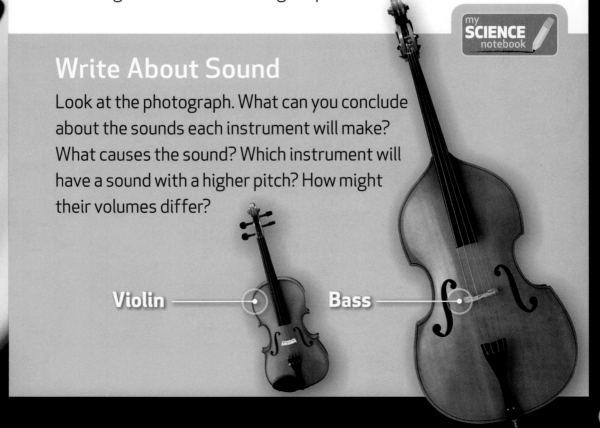

Violin

Bass

PHYSICAL SCIENCE EXPERT: ANIMAL SOUND EXPERT

Dr. William Conner

Dr. William Conner and his team at Wake Forest University have been studying animals and how they use sounds. Bats use sonar to find tiger moths to eat. But tiger moths can actually make high-pitched clicks with their bodies that make the bats' sonar useless!

NG Science: What do you do in your job?

I study animal communication: the signals animals produce, how they travel through the environment, and the messages they give. My students and I have been studying the battle of sounds between insects and bats that prey on them.

NG Science: What are you trying to learn about the sounds animals make?

We know that bats use sound when they track insects to eat. We found that tiger moths can actually answer the signals that bats make by making high-pitched clicking noises. These noises confuse bats. A confused bat cannot catch a moth!

TECHTREK
myNGconnect.com

Digital Library

Bats use their hearing more than they use their sight to find prey.

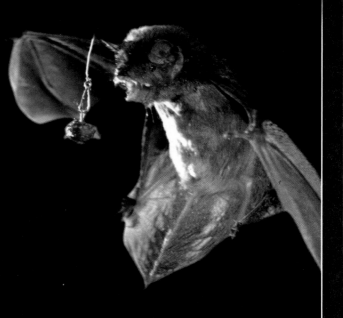

TECHTREK
myNGconnect.com

Student
eEdition

Digital
Library

NG Science: Why is your work important?

Through our work, we've discovered amazing things about how animals adapt to the world. But we've also been thinking about ways this can help people track targets through sound.

NG Science: When did you first know you wanted to study animals and the sounds they make?

One of my teachers in college taught me that nature has amazing things to show students who know where and how to look.

NG Science: What is a typical day in the field like for you?

We work at night! We search for the best spots to set up cameras, microphones, and lights that attract moths. Then we record battles between moths and bats in the nighttime sky.

NG Science: What is your favorite thing about your job?

Seeing and hearing things for the first time makes my job something to look forward to every day. We get to travel to great places like cloud forests and deserts all over the world to make our discoveries.

Special nighttime cameras capture the animals' actions in the dark.

BECOME AN EXPERT

Sound: Using Ultrasound

Can you use **sound** to see? It may seem impossible, but it can be done. Ultrasound uses high-pitched sounds to look inside the human body. The human ear can hear sounds up to 20,000 **vibrations** per second. Ultrasound uses even faster vibrations that have no **volume** . A machine makes vibrations that travel through the body and bounce off organs and bones. This process makes images appear on screens.

ULTRASOUND TIMELINE

1790   **1900**

1790
discovery that bats use their hearing, not their sight, to find their way around

1826
discovery of the speed of sound in water

1914
first working ultrasound machine invented

sound
Sound is a form of energy that you hear.

vibration
A **vibration** is a rapid, back-and-forth movement.

volume
Volume is the level of loudness of a sound.

TECHTREK
myNGconnect.com

Student
eEdition

Digital
Library

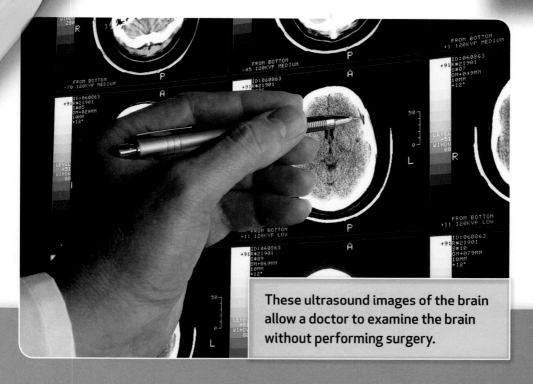

These ultrasound images of the brain allow a doctor to examine the brain without performing surgery.

1930   1940   1950   2000

Late 1930s
first time a doctor used ultrasound to detect brain tumors in humans

Late 1940s
first study of the differences in sound waves as they traveled through the bodies of animals

1950s
first ultrasound detection of twins in a pregnant woman; the patient had to be in water for the ultrasound machine to work

1950s and 1960s
first ultrasound scanner that could be placed directly on the skin of the patient; this was the start of ultrasound as we know it today

Dr. Karl Dussik used ultrasound to examine the human brain.

197

Ultrasound and Medicine In medical ultrasound, vibrations travel through the human body. Ultrasound has been used for many years to detect problems in the body. Tumors, kidney stones, and other masses can be found using ultrasound. Ultrasound is also used for checking the growth and health of unborn babies. The mother's doctor checks the ultrasound pictures carefully to make sure the baby is developing correctly. In addition, the doctor can measure the baby to determine when it will be born.

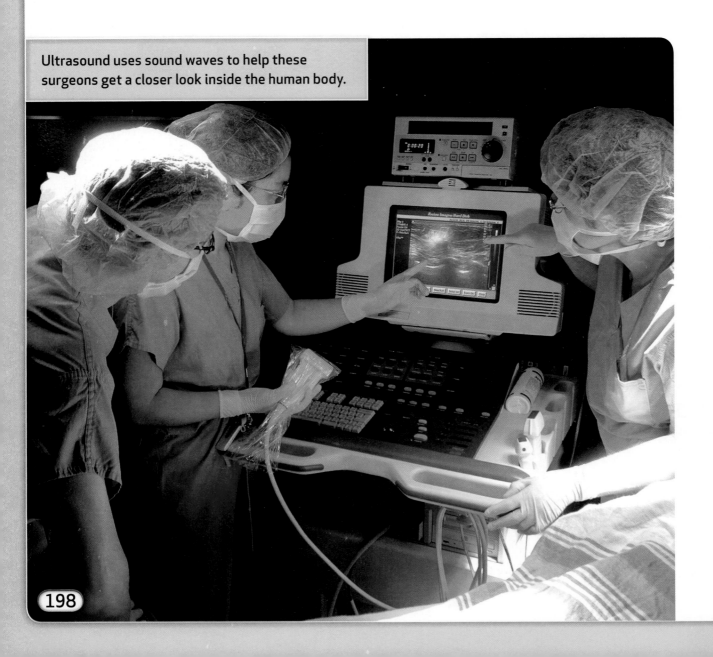

Ultrasound uses sound waves to help these surgeons get a closer look inside the human body.

Another medical use for ultrasound is to monitor heart health. A technician places the ultrasound machine on the patient's chest. Waves of sound travel through the body toward the chest. Once they reach the heart, they echo back to the ultrasound machine. Then the machine uses the data to create a picture. Doctors can see all the parts of the heart without having to do surgery.

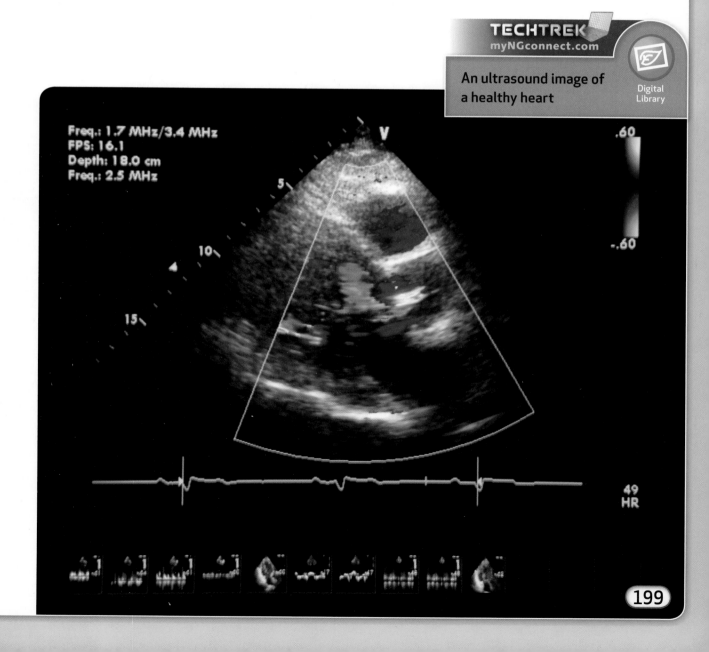

TECHTREK
myNGconnect.com

An ultrasound image of a healthy heart

Digital Library

Ultrasound and Jewelry Gold and silver are both combined with other metals to make jewelry. Some metals get dull, and that means that jewelry needs to be cleaned. Ultrasound keeps jewelry sparkling. In an ultrasound jewelry cleaner, waves travel through a special solution. Like all ultrasound, the sound waves are at a **pitch** that is too high for humans to hear. The waves vibrate in the solution to create bubbles. The bubbles rub against the jewelry, cleaning away dirt.

Ultrasound cleaning keeps precious jewelry sparkling.

pitch
Pitch is how high or low a sound is.

Ultrasound and Sports

Ultrasound can be used to help athletes recover from injuries. It can also help them relax and repair muscles after a hard game or workout. This type of ultrasound is called therapeutic ultrasound. Therapeutic ultrasound uses very fast vibrations that travel through the skin and into the tissue and muscles underneath. The vibration of the sound waves causes the cells of the body to vibrate. This vibration allows muscles to relax and blood to flow.

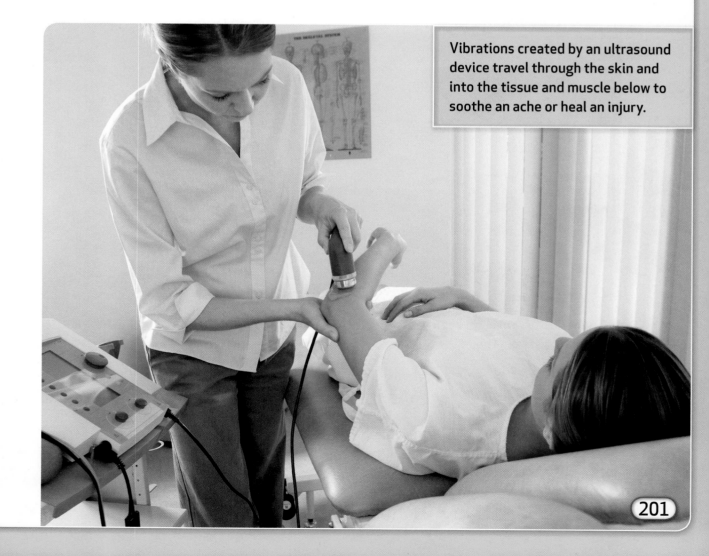

Vibrations created by an ultrasound device travel through the skin and into the tissue and muscle below to soothe an ache or heal an injury.

Ultrasound and Measurement   The word *sonar* combines the words *sound*, *navigation*, and *ranging*. Sonar is a tool that uses sound waves and their echoes to find underwater objects. Sonar operators send a sound wave into the water. When the sound wave meets an object, the wave bounces back as an echo. The sonar operator listens for these echoes. A good sonar operator can tell the difference in echoes. A submarine's echo sounds different from a rock's echo. And a rock's echo sounds different from a whale's. Sonar can tell the distance and direction of an object under the water. Whales and dolphins use a similar type of sonar to find their way, locate food, and communicate.

HOW DOES SONAR WORK?

Sound waves travel through the water. A sonar operator inside the submarine listens and watches a screen that shows images created by the reflected sound waves.

How is ultrasound used for measurement? First, a sonar operator sends out a signal. Part of the signal is reflected as an echo when it hits an object. The operator measures the time it takes for the echo to come back. For example, if it takes four seconds for an echo to come back, the object is two seconds away from the operator. It took the signal two seconds to reach the object and two seconds to return as an echo. The speed of sound in the water is 1,500 meters (about 1 mile) per second. In this case, the object is 3,000 meters (1.9 miles) from the sonar operator.

A sonar operator studies a sonar image to find locations of underwater objects.

CHAPTER
6

SHARE AND COMPARE

Turn and Talk In what ways do people use ultrasound to "see" things? Form a complete answer to this question with a partner.

Read Select two pages in this section. Practice reading the pages. Then read them aloud to a partner. Talk about why the pages are interesting.

my SCIENCE notebook

Write Write a conclusion that tells the important ideas you learned about ultrasound. State what you think is the Big Idea of this section. Share what you wrote with a classmate. Compare your conclusions. Did your classmate make the connection between ultrasound and all of its uses?

my SCIENCE notebook

Draw Draw a picture that shows one of the uses of ultrasound. Include labels in your drawing. Combine your drawing with those of your classmates to make an ultrasound "how to" guide.

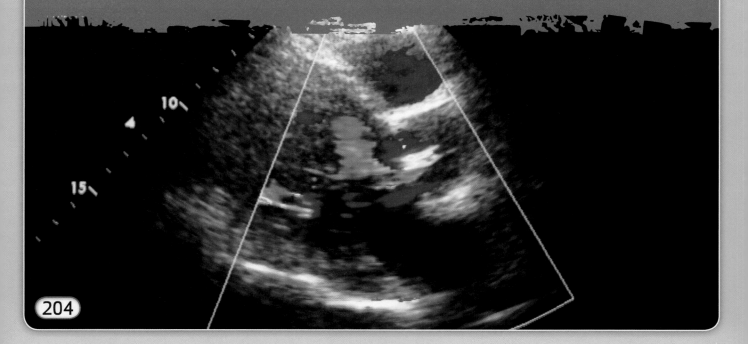

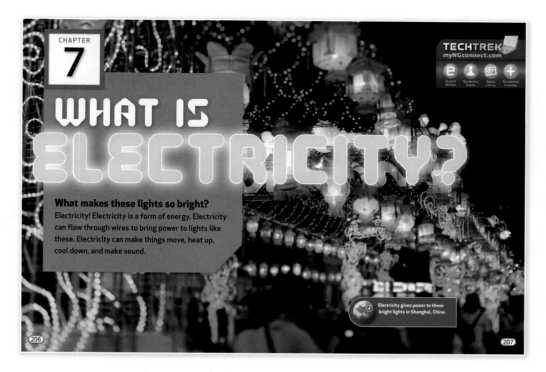

After reading Chapter 7, you will be able to:

- Observe and predict the effects of static electricity on the motion of objects.
  ELECTRICITY

- Demonstrate how electrical energy is transferred and changed through the use of a simple circuit. **ELECTRICITY**

- Classify objects that are good conductors or poor conductors of electricity.
  ELECTRICAL CONDUCTORS AND INSULATORS

- Recognize that a simple circuit must be closed to conduct electricity.
  ELECTRICAL CIRCUITS

- Understand that magnets and electricity produce related forces. **ELECTROMAGNETS**

- Realize that electric charges flowing through a wire can produce a measurable force on magnets and other objects. **ELECTROMAGNETS**

- Understand that electric circuits may produce or use light, heat, sound, and magnetic energy. **ELECTRICAL ENERGY TRANSFORMS**

- Science in a Snap! Observe and predict the effects of static electricity on the motion of objects. **ELECTRICITY**

WHAT IS ELECTR

What makes these lights so bright?

Electricity! Electricity is a form of energy. Electricity can flow through wires to bring power to lights like these. Electricity can make things move, heat up, cool down, and make sound.

TECHTREK
myNGconnect.com

Student
eEdition

Vocabulary
Games

Digital
Library

Enrichment
Activities

ICITY?

Electricity gives power to these
bright lights in Shanghai, China.

SCIENCE VOCABULARY

static electricity
(STA-tik ē-lek-TRIS-it-ē)

Static electricity is a form of electricity in which electrical charges collect on a surface. (p. 210)

This girl's hair is charged with static electricity.

current electricity
(KUR-ent ē-lek-TRIS-it-ē)

Current electricity is a form of electricity in which electrical charges move from one place to another. (p. 213)

Current electricity flows along the wire to move electricity to a lamp so that it lights up.

conductor (Kon-DUK-ter)

A **conductor** is a material through which electricity can flow easily. (p. 214)

Copper is a good conductor of electricity.

my Science Vocabulary

circuit
(SIR-cut)

conductor
(Kon-DUK-ter)

current electricity
(KUR-ent ē-lek-TRIS-it-ē)

electromagnet
(Ē-lek-trō-MAG-net)

insulator
(IN-sū-lā-tur)

static electricity
(STA-tik ē-lek-TRIS-it-ē)

TECHTREK
myNGconnect.com

Vocabulary
Games

insulator (IN-sū-lā-tur)

An **insulator** is a material that slows or stops the flow of electricity. (p. 215)

Plastic is often used as an insulator around wires.

circuit (SIR-cut)

A **circuit** is a path through which controlled electric current flows. (p. 216)

Electricity flows through a simple circuit.

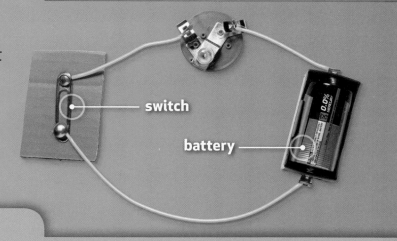

switch

battery

electromagnet
(Ē-lek-trō-MAG-net)

An **electromagnet** is a temporary magnet that is made by an electric current. (p. 220)

When the electric current stops, the electromagnet does not attract objects made of some metal any more.

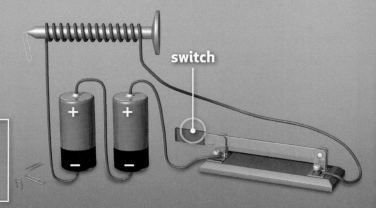

switch

Electricity

Look at the photograph of the girl's hair. What do you think makes her hair stand up that way? Believe it or not, electrical charges are the reason!

An electrical charge is a tiny bit of energy. All matter is made up of particles that have electrical charges.

Suppose you put on socks, shuffled your feet on a carpet, and then touched a metal doorknob. You might feel a small shock. That small shock is caused by the same thing that makes the girl's hair stand up— **static electricity**. Static electricity is a build-up of electrical charges on an object.

Static electricity makes this sock stick to the blanket!

Electrical charges can be either positive or negative. Two objects will pull toward each other, or attract, if one has positive charges and one has negative charges. The balloon in this picture has negative charges. The girl's hair has positive charges. When she holds the balloon near her hair, her hair moves toward the balloon.

Static electricity can be hair-raising!

Look at the balloons below. The arrow shows that they are pushing away from each other. Two objects will repel each other, or push away from one another, if they have like charges. Imagine two balloons have a negative charge. As the balloons are brought together, they repel each other.

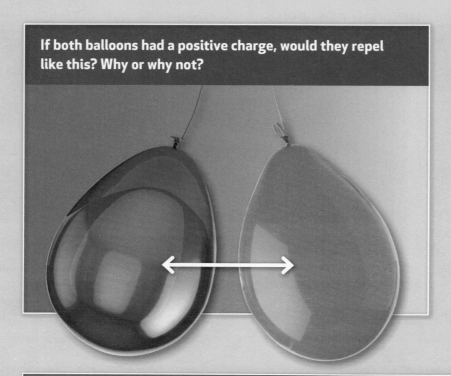

If both balloons had a positive charge, would they repel like this? Why or why not?

Science in a Snap! Observing Static Electricity

Rub a comb on a piece of furry fabric.

Hold the comb near the slowly flowing water.

What happens?

Static electricity can make a small spark. It cannot light a lamp, though. A lamp can be lit by **current electricity**. In current electricity, electric charges move from one place to another. Current electricity can do many things, from lighting a lamp to turning on music to toasting your bread!

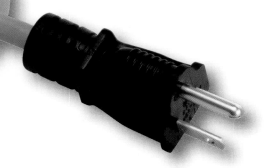

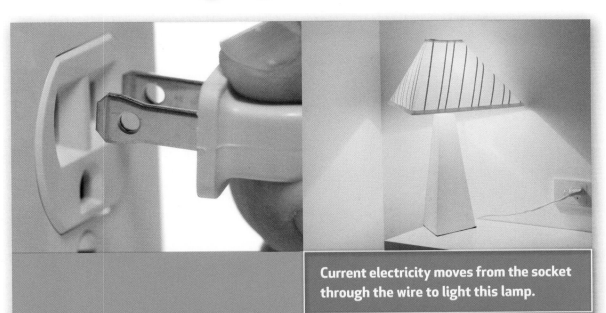

Current electricity moves from the socket through the wire to light this lamp.

Before You Move On

1. What happens when two objects with opposite charges are near each other?
2. How is static electricity different from current electricity?
3. **Draw Conclusions** A balloon is sitting on a table. Another balloon is placed next to it. Neither balloon moves. What can you conclude about the charges of the balloons? Explain.

Electrical Conductors and Insulators

Current electricity brings electricity to homes, schools, and offices. Current electricity can move through wires like those in the picture. Wires are good conductors . Conductors are materials through which electricity can easily flow. The best conductors of electricity are metals, such as silver, copper, gold, and aluminum.

TECHTREK
myNGconnect.com

Digital Library

Electricity flows easily through the copper wires. Copper is a conductor. The plastic around the wires is an insulator. It protects us from the electric current flowing through the wires.

Have you ever noticed that many wires have a material such as plastic or rubber around them? Electricity can be dangerous. That is why wires are covered with materials called insulators. An insulator slows down or stops electricity. Glass, plastic, and rubber are all good insulators.

Electrical towers like these need powerful conductors and insulators to carry electricity and to keep electricity users safe.

Before You Move On

1. What makes a material a good insulator?
2. What are some good conductors of electricity? What makes these materials good conductors?
3. **Evaluate** Why is it important to have conductors and insulators when we use electricity? Explain.

Electrical Circuits

What happens when you switch a flashlight to "on"? A light shines out of the flashlight. What happens on the inside of the flashlight, though? If you open the flashlight, you'll find a **circuit** . A circuit is a path through which electricity flows. The path must be unbroken for electricity to flow through it. When you turn the switch to "on," the circuit is closed and the path is unbroken. When you turn the switch to "off," the path is open, or broken. Electricity stops flowing.

Flashlights use simple circuits and batteries to work.

In some circuits, the electricity can only flow in one path. All the parts of the circuit are in a series, or one part right after another. A circuit like this is called a series circuit. The diagram below shows a series circuit. The electricity can only flow on one path along the wire. If you turn off the switch, the electricity stops flowing.

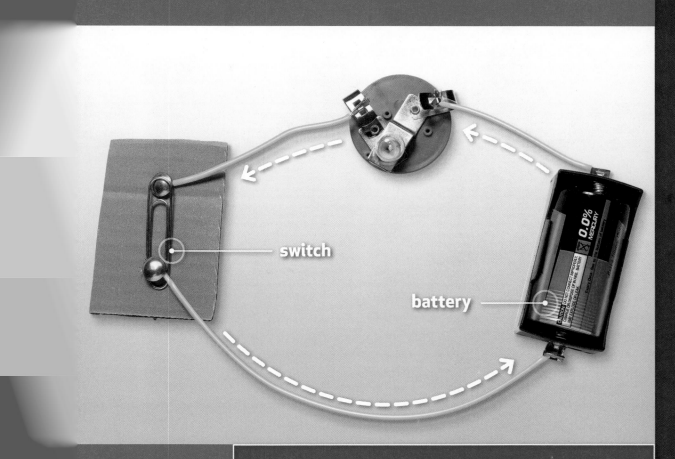

switch

battery

Trace the path in which electricity would flow in the circuit in the picture. This kind of circuit is a series circuit. All the parts of the circuit follow one right after the other. The electricity can only flow on one path through the wire.

Take a look at the string of lights. If the lights were in a series circuit and one light bulb burned out, all the lights would go out. Once the circuit was broken, the electricity would stop flowing. All the lights would turn off.

Parallel circuits are different from series circuits. In a parallel circuit, the electric current can follow more than one path to make a closed circuit. The electrical current splits up among all the paths that are available for it.

These bulbs are in a parallel circuit. Each bulb is part of a complete circuit, so they do not go out when one bulb burns out.

Take a look at the parallel circuit diagram below. The first bulb in the circuit is burned out, but the second bulb stays lit. The electric current has followed more than one path to complete the circuit. You could add additional paths to the circuit. The electricity would simply divide to flow on more paths.

Household devices run on parallel circuits. You can turn off the television, but the light stays on.

In a parallel circuit, when one bulb goes out the other stays lit.

Before You Move On

1. What is a parallel circuit?
2. How is a series circuit different from a parallel circuit?
3. **Analyze** How can you tell if the computers in school are run in a series circuit or a parallel circuit?

Electromagnets

You have probably used a small magnet to pick up small objects such as paperclips. What if you wanted to pick up a giant object, such as a car? People who work with old cars in junkyards need to move the cars from place to place. That's where an **electromagnet** can help. An electromagnet is a coil of wire through which electric current can flow. When the electricity flows, the magnet works. When the electricity stops, the magnet no longer works.

Strong electromagnets make it possible to pick up and move large objects without the objects sticking to the magnets permanently.

Not all electromagnets are giant-sized. You can make a small electromagnet with a nail, some wires, and batteries. The more wire you coil around the nail, the stronger your magnet will be. If you add more batteries, the magnet will be stronger, too.

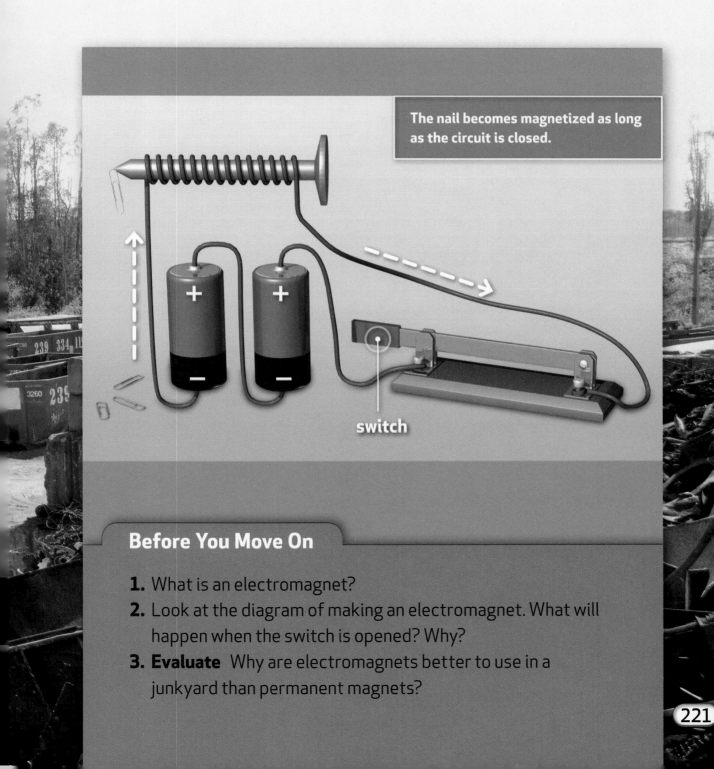

The nail becomes magnetized as long as the circuit is closed.

switch

Before You Move On

1. What is an electromagnet?
2. Look at the diagram of making an electromagnet. What will happen when the switch is opened? Why?
3. **Evaluate** Why are electromagnets better to use in a junkyard than permanent magnets?

Electrical Energy Transforms

You can see bright lights all over the city. The energy to power all these lights comes from electricity. Electrical energy is part of your everyday life. Why is electrical energy so useful? It is so useful because it can transform, or change, into other types of energy: light, heat, motion, and sound.

Electricity lights up Tokyo at night.

Light

If you have ever plugged in a lamp, you have seen electricity transform into light. Devices such as lamps and flashlights use current electricity. Wires are great conductors. They can bring current electricity to objects that convert electricity to useful light.

TECHTREK
myNGconnect.com

Digital Library

Compact fluorescent lightbulbs do not have filaments. Instead, the current electricity moves through a tube of gases. These kinds of bulbs use much less energy than lightbulbs with filaments.

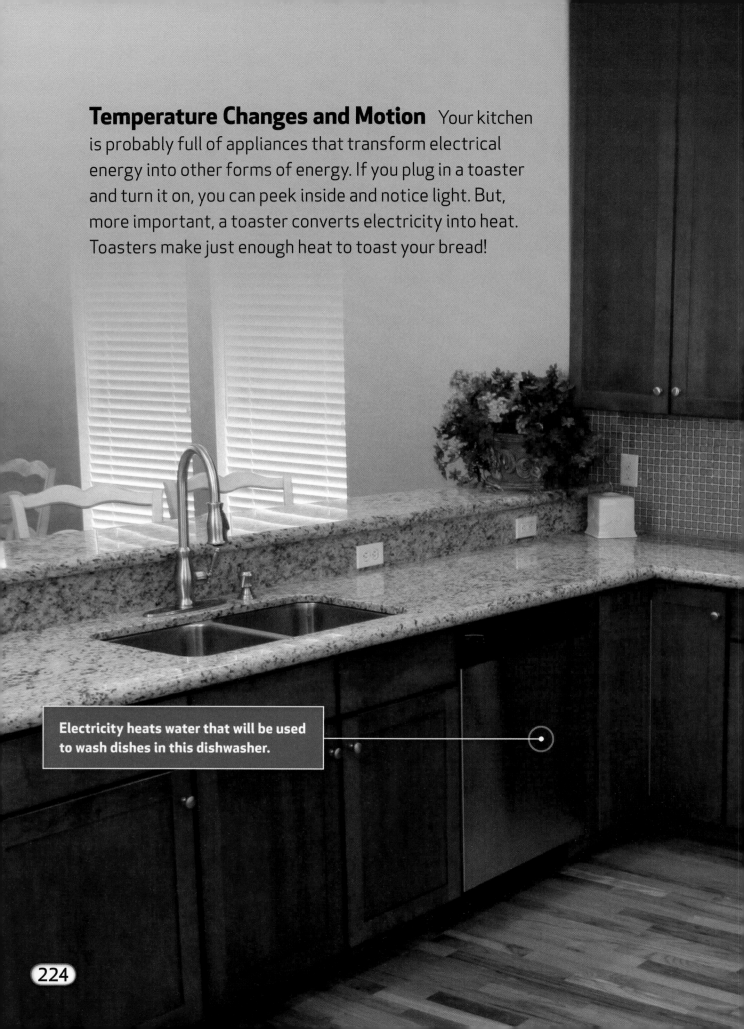

Temperature Changes and Motion Your kitchen is probably full of appliances that transform electrical energy into other forms of energy. If you plug in a toaster and turn it on, you can peek inside and notice light. But, more important, a toaster converts electricity into heat. Toasters make just enough heat to toast your bread!

Electricity heats water that will be used to wash dishes in this dishwasher.

Electrical energy is transformed into motion in your kitchen, too. If you want to mix ingredients, you can mix with a spoon. That takes a lot of energy! If you flip a switch on an electric mixer to "on," electricity powers the appliance and makes the blades turn. You can even control how much electricity flows to the blades to make them turn slowly or quickly.

Electricity transforms to cool your refrigerator. Coils in a refrigerator allow a material in the refrigerator to absorb heat, keeping the refrigerator cool inside.

An electric oven transforms electricity into heat. You can control the amount of electricity that runs through the oven to change the temperature inside the oven.

Sound If you pluck the string of an electric guitar without plugging it in, you can barely hear the sound. Plug it in, though, and the sounds made by the strings become much louder.

TECHTREK
myNGconnect.com

Digital
Library

Electricity transforms into sound to make music!

Electricity can be transformed into sound in two ways. Electricity can cause vibrations that make sound. For example, when lightning strikes, a huge electric charge shoots through the air. It causes the air to expand, making a vibration that you hear as thunder.

Electromagnets can make sound, too. When you press a doorbell, you close a circuit. Current flows through the coils of an electromagnet. The electromagnet attracts a metal arm, which is attached to a hammer. The hammer strikes the bell, making the noise that you hear.

TECHTREK
myNGconnect.com

Enrichment
Activities

When you push the doorbell, you close a circuit. The closed circuit creates the sound you hear.

Before You Move On

1. Reread the title of this lesson. What does it mean?
2. Describe how electricity is turned into motion.
3. **Apply** Name three electrical devices not mentioned in this lesson. For each device, tell what kind of energy electricity is transformed into.

NATIONAL GEOGRAPHIC

TURN ON TO ELECTRIC EELS

Electric eels are long and skinny like eels, but they are not really eels. They are closer relatives to catfish and other "electric" fish.

These famous fish got their names from the fact that they can make a huge electrical charge. Their electricity doesn't come from batteries or wires, though. Electric eels have special cells in their bodies that store power. If they feel danger approaching or want to attack their prey, the cells release the electricity. An electric eel's power is about five times more than the power from a wall outlet.

You would not want to meet an electric eel while you were swimming!

Electric eels live in fresh water in South America. They have to come to the surface to breathe. They have poor eyesight, but they can use electrical charges to light the murky waters ahead of them.

Not many people have been hurt from contact with electric eels. Still, with their huge size and ability to make an electric shock big enough to stun prey, these animals wouldn't be good pets!

Electric eels are long—2 to 3 meters long (6 to 9 feet)—and thin.

All matter has electrical charges. Matter can be neutral, negatively charged, or positively charged. Static electricity is caused by a build up of charge, while current electricity happens when electricity flows. Electricity can be changed to heat, light, motion, or sound. Electricity can also be used to make electromagnets.

Big Idea Electricity is a kind of energy that is caused by the electrical charges of matter. Electricity can be changed into other kinds of energy.

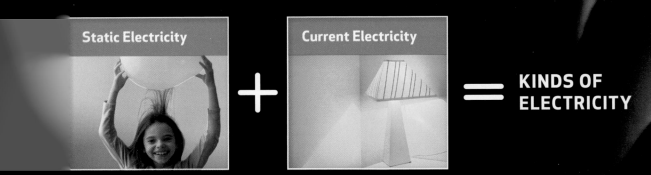

Static Electricity + Current Electricity = KINDS OF ELECTRICITY

Vocabulary Review

Match the following terms with the correct definition.

A. static electricity

B. current electricity

C. conductor

D. insulator

E. circuit

F. electromagnet

1. A material that slows or stops the flow of electricity

2. A temporary magnet that is made by an electric current

3. A material through which electricity can flow easily

4. A form of electricity in which electrical charges move from one place to another

5. A form of electricity in which electrical charges collect on a surface

6. A path through which controlled electrical current flows

Big Idea Review

1. Recall What is an electrical circuit?

2. Describe Write a sentence describing how each of the following devices transforms electrical energy into other forms of energy: toaster, radio, washing machine, lamp.

3. Compare and Contrast How are current and static electricity alike? How are they different?

4. Cause and Effect What causes charged objects to attract?

5. Evaluate Which would you buy: a decorative string of lights that are connected by series circuit or parallel circuit? Why?

6. Infer Suppose you are using an electromagnet to pick up metal pieces from a table. Suddenly all of the pieces fall off the magnet. Why might this have happened?

Write About a Circuit

Explain What is happening in this diagram? What does it show about a circuit?

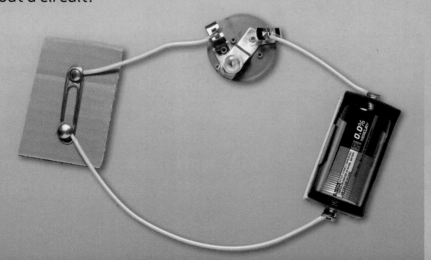

PHYSICAL SCIENCE EXPERT: INVENTOR

How can a backpack make electricity—without plugging in?

Would you like to have an electric backpack? Dr. Larry Rome, a biology professor and researcher, is working on a special backpack that can generate electricity—enough electricity to power an MP3 player, night vision goggles, a global positioning system, and more, all at the same time!

NG Science: What do you study?

Dr. Rome: I am a professor in the Biology Department at the University of Pennsylvania. I study the muscular systems of animals with backbones.

NG Science: What is a Lightning Pack?

Dr. Rome: I invented a special backpack called a "suspended load" backpack. The heavy part of the bag is connected to the frame with a spring or bungee cord that moves up and down while you walk. The motion can be used to create electricity.

NG Science: How did what you study help you design the Lightning Pack?

Dr. Rome: I was teaching about how humans walk and run. I started to wonder, "How can our walking make electricity?"

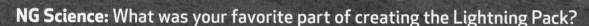

TECHTREK
myNGconnect.com

Student
eEdition

Digital
Library

NG Science: What was your favorite part of creating the Lightning Pack?

Dr. Rome: I had three favorite parts: coming up with the basic idea before I built it, getting the Lightning Pack to work for the first time, and improving my ideas to make a product that people can actually use.

NG Science: What were some tools you used?

Dr. Rome: I used building tools such as drills, but I also had to use treadmills so that people could walk with the backpacks. I used 3-D capture systems, just like in the movies, to study how walking could make electricity.

TECHTREK
myNGconnect.com

Digital
Library

Dr. Rome shows how his lightning pack works.

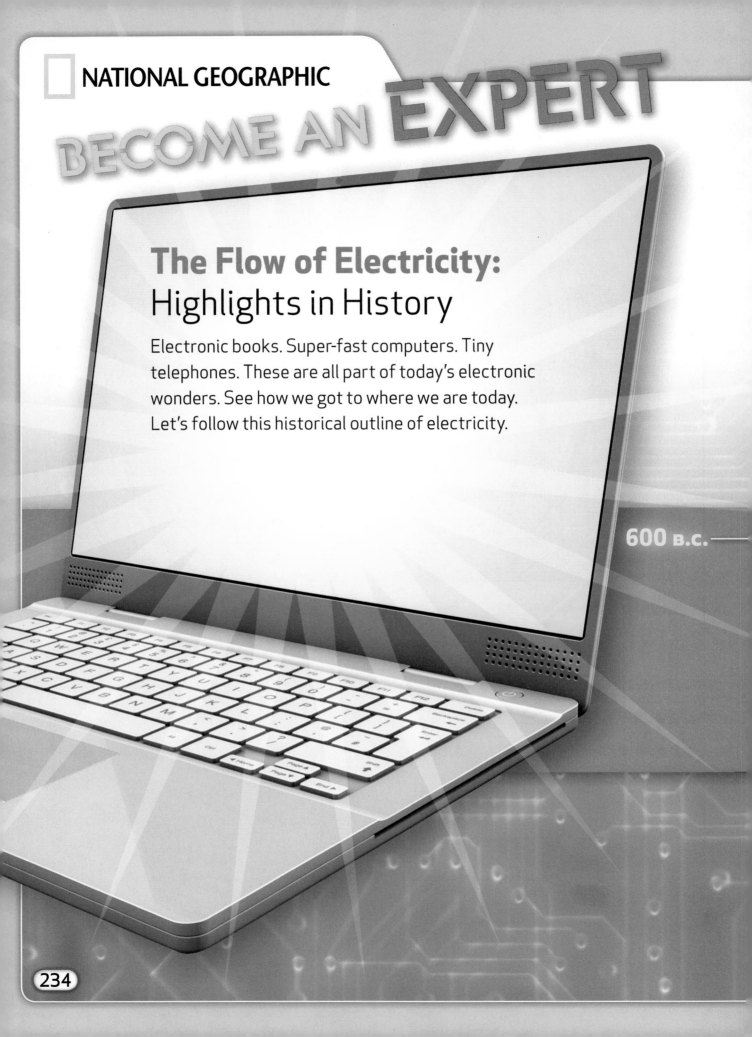

The Flow of Electricity:
Highlights in History

Electronic books. Super-fast computers. Tiny telephones. These are all part of today's electronic wonders. See how we got to where we are today. Let's follow this historical outline of electricity.

600 B.C.

600 B.C. A Greek philosopher (Thales of Miletus) finds that amber becomes charged. He rubs it with silk. Thales is the first person to record ideas about **static electricity** . There is not much progress with electricity until the 17th century.

1600 William Gilbert first uses the term *electricity*. The term comes from the Greek word *elecktron*. *Elecktron* is the Greek word for *amber*.

Thales of Miletus was a Greek philosopher.

1600

static electricity

Static electricity is a form of electricity in which electrical charges collect on a surface.

In addition to being a physicist, Gilbert was also a medical doctor.

1729–1732 Stephen Gray identifies the difference in materials. Some are **conductors**. Others are nonconductors. Today, we also call nonconductors **insulators**.

1729

conductor

A **conductor** is a material through which electricity can flow easily.

insulator

An **insulator** is a material that slows or stops the flow of electricity.

1752 According to many stories, Ben Franklin experiments in a thunderstorm. He ties a key to a kite string. He observes what happens. He discovers that lightning is actually static electricity.

1752

Franklin is believed to have made his discovery about lightning and electricity during a storm in June, 1752.

1800 Alessandro Volta discovers **current electricity** . He experiments with metal disks, wire, fabric, and acid. He finds that by connecting the disks with wire, electrons flow from one disk to another. He uses this stack of disks, fabric, and acid to make the first battery.

1800

Italian physicist Volta shows his electric pile battery.

current electricity
Current electricity is a form of electricity in which electrical charges move from one place to another.

1819–1821 Hans Christian Oersted discovers that an electrical current can move a compass needle. Andre Marie Ampere wants to learn more. He studies this topic. His research leads to Michael Faraday's breakthrough. Faraday invents the **electromagnet** .

Michael Faraday was both a chemist and a physicist.

1819

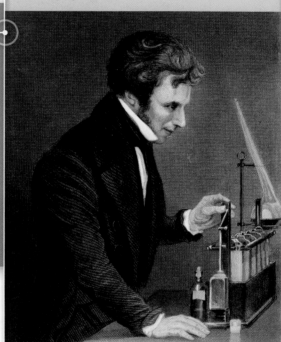

Here is Danish physicist Hans Christian Oersted. He is showing his students how magnetism works.

electromagnet

An **electromagnet** is a temporary magnet made by an electric current.

1844 Samuel Morse invents the telegraph. It uses the basic electrical **circuit** to send sound across wires over long distances. The telegraph is a marvel! It is the first device of its kind.

1844

The telegraph used a system of short and long sounds. Each letter had a symbol of dots and dashes to stand for the sounds.

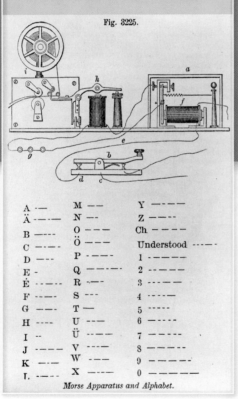

Morse Apparatus and Alphabet.

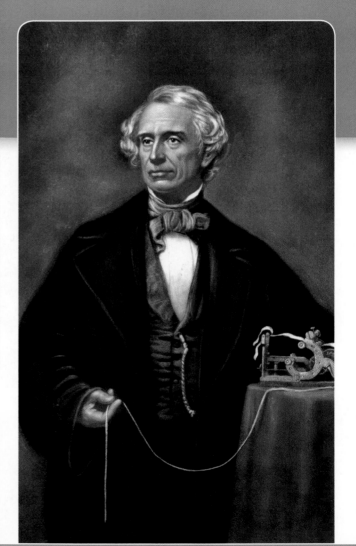

circuit

A **circuit** is a path through which controlled electric current can flow.

1878 Joseph Swan invents the first electric lamp. The lightbulb, however, burns out quickly. Thomas Edison solves that problem by inventing a long-lasting bulb. Edison works with a group of men now called "Edison Pioneers," 28 men who developed the electric light industry. Lewis Latimer makes improvements to some of Edison's original ideas.

Edison is shown here with another one of his famous inventions: the phonograph.

1878

By 1878, Edison and Swan joined together to sell their Ediswan light bulbs.

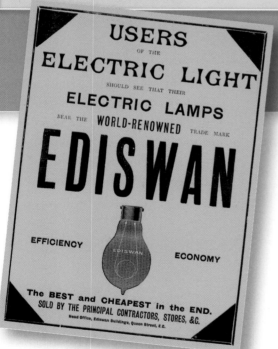

USERS OF THE
ELECTRIC LIGHT
SHOULD SEE THAT THEIR
ELECTRIC LAMPS
BEAR THE WORLD-RENOWNED TRADE MARK
EDISWAN

EFFICIENCY   ECONOMY

The BEST and CHEAPEST in the END.
SOLD BY THE PRINCIPAL CONTRACTORS, STORES, &C.
Head Office, Ediswan Buildings, Queen Street, E.C.

TECHTREK
myNGconnect.com

Digital
Library

Lewis Latimer invented an improved method for making the carbon filaments in lightbulbs.

1902–1913 Electricity is used to make life easier. New electrical appliances come on the scene. In 1902, Willis Carrier makes a crude air conditioner. In about 1908, the portable electric vacuum cleaner is invented. By 1913, the electric refrigerator is invented.

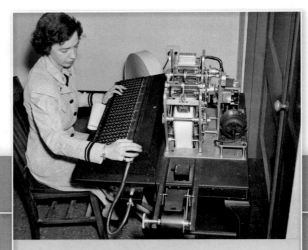

1902

1950

Grace Murray Hopper, a Navy officer, was the first woman computer scientist. She created many early programming languages. She invented the term "debugging" when she found actual moths inside a computer!

The first electric vacuum cleaners were a little bigger and bulkier than vacuum cleaners today!

1950–present In 1953, the first general purpose computer is invented. In 1981, Dr. Adam Osborne invents the first personal computer. Electrical energy is used for many new and exciting devices, such as portable media players. Hybrid cars, available to the public starting in the late 1990s, use energy from both gas and electricity.

Dr. Adam Osborne is shown here with one of his computers.

2007

Electronic books were first available in 2007. The first electronic books could hold 200 titles without illustrations, allowing people to easily carry a wide selection of reading materials!

CHAPTER 7
SHARE AND COMPARE

Turn and Talk How does Joseph Swan's lightbulb compare to Thomas Edison's? Form a complete answer to this question together with a partner.

Read Select two pages in this section. Practice reading the pages. Then read them aloud to a partner. Talk about why the pages are interesting.

 **Write** Write a conclusion that summarizes what you have learned about the history of electricity. State what you think is the Big Idea of this section. Share what you wrote with a classmate. Compare what each of you wrote. Did you recall the same events in the history of electricity?

 **Draw** Form groups of four. Have each person draw a picture that shows an event in the history of electricity. Label your drawing. Put the drawings in the historical order in which they occurred.

Glossary

A

attract (a-TRAKT)
To attract is to pull toward. (p. 116)

C

chemical change (KEM-i-kul CHĀNJ)
A chemical change is a change in which new substances are formed. (p. 56)

chemical energy (KEM-i-kul EN-ur-jē)
Chemical energy is energy that is stored in substances. (p. 158)

One kind of chemical change is decay.

There is chemical energy in the girls' food.

circuit (SIR-cut)
A circuit is a path through which controlled electric current flows. (p. 216)

conductor (kon-DUK-ter)
A conductor is a material through which electricity can flow easily. (p. 214)

current electricity
(KUR-ent ē-lek-TRIS-it-ē)
Current electricity is a form of electricity in which electric charges move from one place to another. (p. 213)

E

electromagnet (ē-lek-trō-MAG-net)
An electromagnet is a temporary magnet that is made by an electric current. (p. 220)

energy (EN-ur-jē)
Energy is the ability to do work or cause a change. (p. 146)

F

force (FORS)
A force is a push or a pull. (p. 82)

friction (FRIK-shun)
Friction is a force that acts when two surfaces rub together. (p. 90)

G

gas (GAS)
A gas is matter that spreads to fill a space. (p. 54)

gravity (GRA-vi-tē)
Earth's gravity is a force that pulls things toward the center of Earth. (p. 94)

H

heat (HĒT)
Heat is the flow of energy from a warmer object to a cooler object. (p. 150)

I

insulator (IN-sū-lā-tur)
An insulator is a material that slows or stops the flow of electricity. (p. 215)

L

light (LĪT)
Light is energy that can be seen. (p. 156)

liquid (LI-kwid)
A liquid is matter that has a definite volume and takes the shape of its container. (p. 52)

M

magnet (MAG-net)
A magnet is an object able to pull some metals toward itself. (p. 114)

This magnet pulls metal objects that contain iron.

magnetic field (mag-NET-ik fēld)
The magnetic field is the area around a magnet where there is a pulling force. (p. 120)

magnetism (MAG-nuh-tism)
Magnetism is a force created by magnets that pulls some metals. (p. 114)

mass (MAS)
Mass is the amount of matter in an object. (p. 22)

matter (MA-ter)
Matter is anything that has mass and takes up space. (p. 10)

mechanical energy (mi-CAN-i-kul E-nur-jē)
The mechanical energy of an object is its stored energy plus its energy of motion. (p. 148)

mixture (MIKS-chur)
A mixture is two or more kinds of matter put together. (p. 18)

This fruit salad is a mixture.

motion (MŌ-shun)
Motion is a change in position. (p. 82)

P

physical change (FI-si-kul CHĀNJ)
A physical change is when matter changes to look different but does not become a new kind of matter. (p. 51)

pitch (PICH)
Pitch is how high or low a sound is. (p. 186)

These guitars can make sounds with many different pitches.

pole (PŌL)
The pole is the part of the magnet where the force is strongest. (p. 116)

property (PROP-er-tē)
A property is something about an object that you can observe with your senses. (p. 10)

R

repel (ra-PEL)
To repel is to push away. (p. 116)

EM3

S

solid (SO-lid)
A solid is matter that has a definite shape and volume. (p. 50)

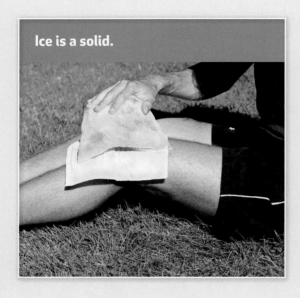
Ice is a solid.

solution (so-LŪ-shun)
A solution is a mixture of two or more kinds of matter evenly spread out. (p. 21)

sound (SOWND)
Sound is a form of energy that you hear. (p. 178)

speed (SPĒD)
Speed is the distance an object moves in a period of time. (p. 88)

states of matter (STĀTS UV MA-ter)
States of matter are the forms in which a material can exist. (p. 50)

static electricity (STA-tik ē-lek-TRIS-it-ē)
Static electricity is a form of electricity in which electrical charges collect on a surface. (p. 210)

V

vibration (vi-BRĀ-shun)
A vibration is a rapid, back-and-forth movement. (p. 178)

volume (VOL-yum)
Volume is the amount of space something takes up. (p. 26)

volume (VOL-yum)
Volume is the level of loudness of a sound. (p. 182)

This juice is a solution of drink mix and water.

Index

Credits

Front Matter

About the Cover (bg) Roger Bamber/Alamy Images. (t, inset) Roger Bamber/Alamy Images. (b, inset) Dave Porter/Alamy Images. **Back Cover** (bg) Roger Bamber/Alamy Images. (tl) Calit2. (tr) Erik Jepsen. (c) Erik Jepsen. (bl) Ted Wood/Aurora/Getty Images. (br) James L. Stanfield/National Geographic Image Collection. **ii-iii** Jim Foley/Getty Images. **iv-v** Allan Baxter/Getty Images. **vi-vii** Mikael Damkier/Shutterstock. **vii** blickwinkel/Alamy Images. **viii-ix** Thomas Del Brase/Getty Images. **ix** Todd Korol/Getty Images. **x-1** Paul Thompson/Corbis. **2** (t) Thomas Owen Jenkins/Shutterstock. (c) David Chasey/Photodisc/Jupiterimages. (b) John Foxx Images/Imagestate. **2-3** Jim Lozouski/Shutterstock. **3** (t) Chris Rady/Jupiterimages. (tc) Jim West/Alamy Images. (bc) Karl Ammann/Getty Images. (b) imageshunter/Shutterstock. **4** (bg) Erik Jepsen. (inset) Erik Jepsen.

Chapter 1

5, 6-7 Thomas Owen Jenkins/Shutterstock. **8** (t) Kendra Nielsam/Shutterstock. (c) Jan Butchofsky/Corbis. (b) Ivan Bajic/iStockphoto. **9** (t) Evgeny Kuklev/iStockphoto. (c) E.R. Degginger/Color-Pic Inc. (b) Suk-Heui Park/Getty Images. **10-11** Jan Butchofsky/Corbis. **11** (tl) Barbro Bergfeldt/Shutterstock. (tr) Rtimages/Alamy Images. (b) Kendra Nielsam/Shutterstock. **12** Daniel Kerek/Alamy Images. **13** (bg) TravelStockCollection - Homer Sykes/Alamy Images. (tl) Andrew Northrup. (tr) Andrew Northrup. **14** (tl) Elena Schweitzer/Shutterstock. (tr) Thinkstock Images/Getty Images. (cl) Simon Krzic/Shutterstock. (cl) Bershadsky Yuri/Shutterstock. (cr) Michael Cavén/iStockphoto. (cr) John Foxx Images/Imagestate. (bl) Sabine Scheckel/Getty Images. (br) iStockphoto. **14-15** Kevin Fleming/Corbis. **15** (l) Nguyen Thai/Shutterstock. (r) MetaTools. **16** Hampton-Brown/National Geographic School Publishing. **17** (t) PhotoDisc/Getty Images. (c) Igor Karon/iStockphoto. (b) Don Smith/Getty Images. **18** Ivan Bajic/iStockphoto. **19** (t) J.Garcia/photocuisine/Corbis. (b) Lori Sparkia/Shutterstock. **20** Viorika Prikhodko/iStockphoto. **20-21** Evgeny Kuklev/iStockphoto. **21** Dan Brandenburg/iStockphoto. **22** PhotoDisc/Getty Images. **23** (t) E.R. Degginger/Color-Pic Inc. (b) Jeremy Woodhouse/Blend Images/Corbis. **24** (l) Hampton-Brown/National Geographic School Publishing. (r) Hampton-Brown/National Geographic School Publishing. **25** Hampton-Brown/National Geographic School Publishing. **26** Serdar Yagci/iStockphoto. **26-27** Suk-Heui Park/Getty Images. **27** Grant Heilman Photography/Alamy Images. **28** Ieva Geneviciene/Shutterstock. **29** (l) Hampton-Brown/National Geographic School Publishing. (r) Hampton-Brown/National Geographic School Publishing.
30 Ned Therrien/Visuals Unlimited. **30-31** (t) Dorling Kindersley/Getty Images. (b) Tim Laman/National Geographic Image Collection. **31** Steve Smith/Getty Images. **32** (l) Rtimages/Alamy Images. (c) Nguyen Thai/Shutterstock. (r) Thinkstock Images/Getty Images. **32-33** altiso/Shutterstock. **33** Hampton-Brown/National Geographic School Publishing. **34** Albert Yu-Min Lin, Ph.D. **34-35** Norman Chan/Shutterstock. **35** (l) Filip Fuxa/Shutterstock. (r) DeshaCAM/Shutterstock. **36** Ira Block/National Geographic Image Collection. **36-37** adrian arbib/Alamy Images. **38** MODIS Science Team/NASA - Visible Earth. **38-39** Stockbyte/Getty Images. **39** Albert Yu-Min Lin, Ph.D. **40** (t) Albert Yu-Min Lin, Ph.D. (b) Albert Yu-Min Lin, Ph.D. **40-41** Albert Yu-Min Lin, Ph.D. **41** Albert Yu-Min Lin, Ph.D. **42** Erik Jepsen. **43** Albert Yu-Min Lin, Ph.D. **44** Erik Jepsen.

Chapter 2

45, 46-47 David Chasey/Photodisc/Jupiterimages. **48** (t) Photodisc/Angelo Cavalli/Getty Images. (c) Jeannot Olivet/iStockphoto. (b) Losevsky Pavel/Shutterstock. **49** (t) Lucian Coman/Shutterstock. (c) ZTS/Shutterstock. (b) Tom Pantages. **50-51** Losevsky Pavel/Shutterstock. **51** Jeannot Olivet/iStockphoto. **52** Lucian Coman/Shutterstock. **52-53** Photodisc/Angelo Cavalli/Getty Images. **54-55** David Partington/iStockphoto. **55** ZTS/Shutterstock. **56** Tom Pantages. **57** Timothy Goodwin/iStockphoto. **58-59** Matt Antonino/Shutterstock. **59** (t) Maxim Tupikov/Shutterstock. (b) Anne Clark/iStockphoto. **60** (l) Artville. (r) PhotoDisc/Getty Images. **60-61** George Doyle/Getty Images. **61** (l) Andrew Northrup. (r) Andrew Northrup. **62** WaterFrame/Alamy Images. **62-63** Emory Kristof/National Geographic Image Collection. **64** (l) Timothy Goodwin/iStockphoto. (cl) Matt Antonino/Shutterstock. (cr) Anne Clark/iStockphoto. (r) George Doyle/Getty Images. **64-65** Stockbyte/Getty Images. **65** Jeff Chiasson/iStockphoto. **66** Barbara White. **67** (bg) Daisuke Morita/Getty Images. (l) Barbara White. (r) Susumu Nishinaga/Photo Researchers, Inc. **68** Dagli Orti /The Art Archive. **68-69** (t) Shirly Friedman/iStockphoto. (b) Jason Edwards/National Geographic Image Collection. **69** Victoria & Albert Museum, London/Art Resource, Inc. **70** (bg) Bill Varie/Corbis. (inset) PhotoDisc/Getty Images. **71** (l) Tony Freeman/PhotoEdit. (r) David R. Frazier Photolibrary, Inc./Alamy Images. **72** Lourens Smak/Alamy Images. **73** Tatiana Markow/Sygma/Corbis. **74** (t) John Zoiner/Getty Images. (b) Semen Lixodeev/Shutterstock. **75** Stockbyte/PunchStock. **76** Jason Edwards/National Geographic Image Collection.

Chapter 3

77, 78-79 Gordon Wiltsie/National Geographic Image Collection. **80** (t) Compassionate Eye Foundation/Jetta Productions/Getty Images. (b) First Light/Alamy Images. **81** (t) Image Source/Getty Images. (c) Tatiana Morozova/Shutterstock. (b) Digital Vision/Getty Images. **82-83** Compassionate Eye Foundation/Jetta Productions/Getty Images. **83** First Light/Alamy Images. **84** blue jean images/Getty Images. **85** (l) muzsy/Shutterstock. (r) muzsy/Shutterstock. **86-87** Jorge Royan/World of Stock. **87** (l) Wendy Hope/Jupiterimages. (r) Steve Allen/Photo Researchers, Inc. **88-89** Image Source/Getty Images. **90** (l) Andrew Northrup. (r) Andrew Northrup. **90-91** Tim Laman/National Geographic Image Collection. **91** Tatiana Morozova/Shutterstock. **92** Hampton-Brown/National Geographic School Publishing. **92-93** (bg) Hampton-Brown/National Geographic School Publishing. **93** Hampton-Brown/National Geographic School Publishing. (inset) Hampton-Brown/National Geographic School Publishing. **94-95** Digital Vision/Getty Images. **95** Hampton-Brown/National Geographic School Publishing. **96-97** R. S. Ryan/Shutterstock. **97** David Toase/Getty Images. **98** (l) Compassionate Eye Foundation/Jetta Productions/Getty Images. (c) Tim Laman/National Geographic Image Collection. (r) Digital Vision/Getty Images. **98-99** Seth Joel/Getty Images. **99** Chris Johns/National Geographic Image Collection. **100** (t) Cynthia Emerick. (b) Cynthia Emerick. **101** Cynthia Emerick. **102** Mike Powell/Getty Images. **102-103** Image Source/Jupiterimages. **103** (l) Markus Boesch/Getty Images. (r) David Madison/Getty Images. **104** Stockbyte/Jupiterimages. **105** David Madison/Jupiterimages. **106** (t) Sports Illustrated/Getty Images. (b) Doug Pensinger/Getty Images. **106-107** David Madison/Getty Images. **108** David Madison/Getty Images.

Chapter 4

109, 110-111 Mark Lennihan/AP Images. **112** (t) Mike Kemp/Getty Images. (c) Digital Vision/Alamy Images. **113** (b) Ted Foxx/Alamy Images. **114** Digital Vision/Alamy Images. **115** Mike Kemp/Getty Images. **117** (bg) Stuart Dee/Getty Images. (inset) m/iStockphoto. **118** (t), (br) PhotoDisc/Getty Images. (bl) Goran Kuzmanovski/Shutterstock. (cl) Andrew Northrup. (cr) Andrew Northrup. **119** (tl) Stefan Ataman/iStockphoto. (tr) PhotoDisc/Getty Images. (cl) Martin Smith/Shutterstock. (cr) Denis Pogostin/iStockphoto. (bl) Stockbyte/Getty Images. (br) PhotoDisc/Getty Images. **121** Ted Foxx/Alamy Images. **122** Ted Foxx/Alamy Images. **123** Yamada Taro/Getty Images. **124** (t) Luca di Filippo/iStockphoto. (c) iStockphoto. (b) Scientifica/Visuals Unlimited/Alamy Images. **125** Chris Rady/Jupiterimages. **126-127** Photodisc/Getty Images. **127** Mark Thiessen/Hampton-Brown/National Geographic School Publishing.

128 (l) Gustoimages/Photo Researchers, Inc. (r) Lester Lefkowitz/Corbis. **129** (t) Dr. Vivian Lee. (b) Olaf Doering/Alamy Images. **130-131** Peter Pinnock/Getty Images. **131** Gallo Images-Anthony Bannister/Getty Images. **132** Herbert Zettl/Corbis. **132-133** Michael Fay/National Geographic Image Collection. **133** Pal Hermansen/Getty Images. **134** Michael Durham/Minden Pictures/National Geographic Image Collection. **135** Michael Durham/Minden Pictures/National Geographic Image Collection. **136** Jeffrey L. Rotman/Corbis. **137** Peter Johnson/Corbis. **138** imagebroker/Alamy Images. **138-139** Karen & Ian Stewart/Alamy Images. **139** Bates Littlehales/National Geographic Image Collection. **140** Peter Pinnock/Getty Images.

Chapter 5

141, 142-143 Todd Korol/Getty Images. **144** (t) Corbis/Jupiterimages. (c) Hampton-Brown/National Geographic School Publishing. (b) Richard Nowitz/National Geographic Image Collection. **145** (t) Victor R. Boswell, Jr./National Geographic Image Collection. (b) Bounce/Getty Images. **146-147** Corbis/Jupiterimages. **148-149** Hampton-Brown/National Geographic School Publishing. **149** Hampton-Brown/National Geographic School Publishing. **150-151** Richard Nowitz/National Geographic Image Collection. **151** (l) Andrew Northrup. (r) Andrew Northrup. **152** J. Baylor Roberts/National Geographic Image Collection. **153** Arpad Benedek/iStockphoto. **154** (t) C Squared Studios/Photodisc/Getty Images. (c) Chuck Swartzell/Visuals Unlimited. (b) ricardo azoury/iStockphoto. **155** (bg) Thomas Del Brase/Getty Images. (inset) simon edwards/iStockphoto. **156-157** Victor R. Boswell, Jr./National Geographic Image Collection. **157** Hampton-Brown/National Geographic School Publishing. **158-159** Bounce/Getty Images. **159** Image Source/Alamy Images. **160** AP Images. **160-161** Jim West/Alamy Images. **162** (l) Hampton-Brown/National Geographic School Publishing. (cl) Richard Nowitz/National Geographic Image Collection. (cr) Victor R. Boswell, Jr./National Geographic Image Collection. (r) Bounce/Getty Images. **162-163** Erik Isakson/Getty Images. **163** Jutta Klee/Getty Images. **164** Allison Gray. **165** Jim West/Alamy Images. **166** Harald Sund/Getty Images. **167** Charles Palek/Animals Animals. **169** U.S. Bureau of Reclamation/Photo Researchers, Inc. **170-171** Michael T. Sedam/Corbis. **172** Michael T. Sedam/Corbis.

Chapter 6

173, 174-175 Karl Ammann/Getty Images. **176** (t) Patrick Byrd/Alamy Images. (b) Matej Michelizza/iStockphoto. **177** (t) Hanka B./Shutterstock. (b) Plush Studios/Getty Images. **178** Jason Stitt/Shutterstock. **178-179** (t) Matej Michelizza/iStockphoto. (b) Patrick Byrd/Alamy Images. **179** Alistair Forrester Shankie/iStockphoto.